Seeker
freedom

Reclaiming Feminine Wisdom

A Memoir
By
Yvonne Winkler

Freedom Seeker Copyright © 2022 Yvonne Winkler

All rights reserved. No part of this book may be reproduced by any mechanical, photographic, or electronic process, or in the form of phonographic recording; nor may it be stored in a retrieval system, transmitted, or otherwise copied for public or private use without the prior written permission of the publisher.

Lotus Consulting Inc.
Calgary, AB, Canada
info@yvonnewinkler.com

Cover art: KP Design

ISBN 978-1-7386967-0-3eBook
ISBN 978-1-7386967-1-0Paperback

First Edition

This is a work of nonfiction. Some names and identifying details have been changed.

To all the women who believe they're not enough.

1
COMPROMISE

What have I done? I sat on my parents' couch in an established southwest Calgary neighbourhood and stared out the window into the rain-soaked March morning with a twist in my stomach. As I blew my nose for the umpteenth time that day, I kept wondering, *did* I really just leave my perfect life, security, and promising career to go back to *uncertainty, a shoestring budget, and what little I could fit into a forty pound rucksack?*

Here I was thirty-two years old, single again, homeless, and, as of the fifteenth of March, with no more paycheck deposits. While most of my friends were getting married, buying houses and starting families, I was willing to walk away from a decade of career building, a cozy downtown Vancouver apartment, friends, health and dental coverage, and all my other earthly possessions. It was impulsive and illogical and several people, especially my boss, warned me about leaving great opportunities behind. But my longing for a fresh start, new places and faces was so much deeper than any fear of missing out on the next promotion or real estate transaction.

The truth was, I didn't know what I wanted anymore in a relationship, in a job or in my life. I was so tired chasing after hefty career goals, titles, and the recognition I thought would buy me the freedom and independence my parents came to this country for. Now, I just wanted to tear off the shackles that held me in the confines of social expectations, job security and certainty. For the first time ever I chose me and listened to what my heart wanted,

5

instead of doing what I thought I needed to in order to be accepted as an upstanding member of society.

I diligently spent months planning my adventure. I read books on how to be safe as a female traveler and gathered the necessities of solo backpacking through Europe for however long it would take to find whatever I was looking for. I made lists of places I wanted to see, careful to not plan my trip in too much detail. I was craving freedom from any rules or limitations. There was a wildness emerging inside of me that was exhilarating and intoxicating. I wanted to be in complete control over my life and decisions, including if or when I'd come back. That was the main reason I chose not to take the recommended sabbatical.

Focused upon a raindrop suspended on the windowpane, I rewound the tape of the last six years of ambitious career ladder climbing in my mind. After a painful separation from my then fiancé, crippling student loan debt, and zero friends in this new city, I landed a position as an account manager with a large Canadian insurance and investment brokerage where my previous experience as a door-to-door insurance saleswoman gave me an edge with the brokers. It was an entry-level administrative position that entailed managing the full-cycle life insurance applications from submission to commission, and it was a heck of a lot easier than traveling across the prairies, knocking on doors, selling funeral insurance. After a couple of years on the road, sitting at strangers' kitchen tables, working seven days a week, I enjoyed the routine of the forty-hour work week with weekends at home, doing what twenty-five-year-olds do for a small but consistent paycheck. And, as my new colleague liked to say, it sounds impressive when you point out that you work for a downtown Calgary financial company. Emphasis being on downtown.

It was a welcome lift to my professional ego. Having lost not only my relationship but also the delusion I would earn a six-figure income right out of university, the tension in my jaw, neck, and shoulders began to release when, on the first and fifteenth of every month, a consistent sum of money was deposited into my bank account.

When an opportunity on the marketing team became available I immediately applied because from there I saw a greater chance for advancement into leadership. I landed the position as the new marketing assistant where, instead of processing paperwork in a hallway cubicle, I got my own office, a chance to advance my professional skills and, as an added bonus, my own reserved underground parking spot. I also broke the $50,000 per year income barrier which, according to my new boyfriend, was a great career milestone.

This was my first taste at corporate success, and I was hooked. I intended to make the most of this opportunity, build relationships and do everything I could to hopefully, one day, take my seat at the head of the boardroom table.

Finally, four years after walking across the stage to accept my business diploma, I made enough to pay my rent, buy groceries, and have money left over to treat myself to a new pant suit and a pair of purple BCBG pumps, which were on sale. I also quickly discovered that this kind of privilege came at the price of undefined boundaries, mostly in the form of subjugation and objectification.

Sitting in my office going through my emails, I noticed an invitation to the Calgary Stampede, world famous as the largest outdoor rodeo, when my manager walked in and I asked,

"Drinking vodka at breakfast?" I looked at my manager baffled. "And I'm getting paid to do this?"

For most Alberta businesses, this was a huge sponsorship affair that ate a big chunk of the annual marketing budget. As an attractive, young, blond woman I was sent to represent our organization and show our clients a yahoo-good time at what was known as Bullshooters Breakfast, a fundraiser shindig that started at eight o'clock in the morning and went until noon officially, and into the night unofficially.

"Yes," he replied with a grin. "I invited our best advisers, and our insurance partners take turns to buy drink tickets. I'll have some for you to give out to the brokers. Oh, and the trick is to order orange juice. That way they all assume you're drinking vodka orange." And with that he turned on his heel and left my office.

Watching what free drinks and the Wild West did to the masses during Stampede was a standing joke, and more of a show than the bucking bulls at the rodeo. Once I attended the event and witnessed firsthand male clients suddenly forgetting that they had wives at home and bluntly proposing more than a friendly two-step, I knew this was beyond what I was willing to do for my job. It was the most disheartening event of my career, a real eye-opener of what it meant to operate as a young, driven woman in a world built by men, for men.

Nevertheless, I advanced quickly, putting in countless hours of unpaid overtime because I thought that was what was required to climb the ranks. I even began the tedious path of becoming a certified financial planner (CFP), the standard for the financial planning profession worldwide, and surely my best option to gain respect and credibility amongst my peers and our clients.

The young, dynamic woman leading our company dressed the part and walked the walk. I was mesmerized by how she commanded the room with elegance and confidence, and I wanted to be respected and revered like that. Having had a strong sense for inequality from early on in my life, seeing women in business and at the top gave me hope for where I was headed, and making six figures by the time I reached thirty was a goal to me like having two kids and a white picket fence was to others. And so, I began the rigorous education program to obtain the CFP designation.

My hard work began to show some payoffs. In just two short years I had established a solid reputation as a dedicated employee with excellent client management skills, and my name got tossed around other executive teams as a strong contender for bigger opportunities. With options by my side, I began to shift my focus towards improving my personal life.

My boyfriend and I spent every weekend hiking in the Rockies, rollerblading the paths along the Bow River or biking through Fish Creek Park. In the two years we had been dating he helped me gain a real appreciation for living close to the mountains, and now I was craving more. I wanted to be in a place where I could hike, bike, ski, and watch the sun set over the ocean, possibly on the same day. He suggested Vancouver, a city on the Pacific Ocean and nestled into the North Shore mountains. I had never been there but it sounded idyllic, so I asked my manager if they had my current position or better available. They didn't. But because I was considered an asset to the company and being groomed by the CEO to become a manager, they happily cleared an office for me anyway. Three short months later the moving truck pulled up to our new inner-city condo, complete with rooftop pool and 360-degree panoramic view of downtown Vancouver, only a few steps from my new office, new title, and a small raise.

I had to pinch myself to know I wasn't dreaming as I stood on my little balcony overlooking Coal Harbour. I took a deep breath of the salty, humid air laced with a hint of sweet magnolia which draped the sidewalks to Vancouver's famous seawall. This was my new life. The vibrant birds-of-paradise, sexy pink heliconias, and oriental lilies enticed me to a more extravagant weekly grocery shopping experience and awoke my inner chef wanting to host lavish gourmet dinner parties. Saturday mornings turned into hour long coffee and talk time with my friends back in Calgary, watching the sea planes take off and land skillfully on a small strip of water, and the paddle wheeler leave for its first harbor tour. I soaked in the warm oceanic climate and the lush mountain scenery every chance I got.

Something about the athleticism the people held here made me want to be healthier. I picked up running as a form of stress relief, embraced the challenge of the Grouse Grind trail, learned the most important sea kayaking skills, explored Whistler on foot and on skis, snowshoed Cypress Mountain in the dark, and every other Sunday, after a long walk around the seawall, I met my friends for brunch at The Elbow Room Cafe. It appeared that my life was complete. I was reigning my domain from the top of the world. But a feeling that I was missing something, something important, continued to bring me down and made it difficult to put on my suit and go to work.

I thought maybe it was because I had trouble finding my place in the office. For one, the Vancouver marketing department was twice the size of the Calgary team, fostering a more competitive rather than collaborative environment. Vancouver was the third largest city in Canada and as such, far denser and more culturally diverse than the western cattle town I had called home for the past seven years. There was the stress of appearances, both physical and economic.

With a cup of freshly brewed coffee in hand, I made my way from my tiny office down the narrow hallway into the windowless, florescent lit boardroom that, much like the rest of the office, lacked some serious TLC and updating. Due to limited space the boardroom also served as storage for themed party decorations, an attempt to bring the departments together over occasional potluck lunches. Extra chairs were stacked up against the khaki-colored wall that looked as washed-up as the mug I placed on the racetrack shaped laminate table. While the other marketing members leisurely strolled in, I stared at the shadow on the wall across from me and wondered what inspirational quote might have hung there once and why it was taken down, when my thoughts were interrupted with a loud voice.

"How was your weekend, Winkler?" Kevin, our advanced marketing sales director, called everyone by their last name as a form of endearment.

Freedom Seeker | 11

"Good. Thanks," I replied guarded. I didn't feel comfortable divulging too much personal information within earshot of my manager.

The six of us cramped around the table with our notepads and the guys exchanged a few hockey updates before our VP walked in sharply and everyone fell quiet. The regional VP, a middle-aged woman of slim build and stiff posture, closely watched my every move like she was trying to catch me doing something wrong. The Monday morning meetings were dominated by a threat that if we didn't meet our sales quotas by the end of the month we'd be replaced. This finger wagging, authoritative leadership style was not working for me or the others. The fact that she couldn't figure that out frustrated me and left me with no admiration for her position.

After the usual debrief of our activities from the past week, the chief marketing officer, an uptight middle-aged man with an inflated sense of importance, walked into our meeting and sat down next to the VP at the head of the table to discuss ideas for the annual broker appreciation event. Kevin immediately piped up and suggested golf, followed by more sporting events like hockey games, football and such. Not only could I not have cared less about golf or other impact sports, I was sure that the few female advisors we had would feel the intended appreciation more in a neutral setting.

"What about a dinner cruise?" I exclaimed with a big smile, but nobody registered the suggestion and they continued to blurt out more stuff the guys really wanted to go do; poker night at the casino, Scotch tasting, renting out a theater.

"What about a dinner cruise through English Bay during the festival of lights?" the CMO suggested.

"That's a fantastic idea," a unison chant erupts around the table.

I gripped the armrest of the worn out round upholstered chair to prevent me from leaping out of it. *Of course, he gets all the accolades.* I thought, moping. I had been busting my ass and nobody saw me for more than the cute, young, blond chick. My ideas weren't acknowledged or fostered like I was used to with my old team, and I sorely missed the comfort of an established reputation. But my biggest issue was that I accepted the first offer that came my way, despite working for $0.73 to the dollar my male colleagues earned. None of this was as obvious to me in Calgary as it became here, in the big city, where things were done differently.

Every day felt like a battle for survival and four thirty couldn't come fast enough. My enthusiasm for my job died at the equal rate as my ability to bring new ideas to the monthly marketing huddles. Not knowing how else to compete with five other marketing specialists, all male, and two women who ranked superior to me, I booked that seven-hour final exam to become a certified financial planner, a title I hoped would earn me equal pay and maybe open new doors.

To add to my misery, I discovered over Thanksgiving dinner, that my latest three-year relationship had been built on one giant pathological lie and with the blessing of a therapist who assured me that there was little I could do to change that, I moved into an eight hundred square foot bachelor suite on the opposite side of downtown.

It had been eight months since I arrived in Vancouver, and I was noticeably withering instead of flourishing. During one of my Saturday morning walks around the seawall, it suddenly occurred to me that no matter how hard I worked, what designation came after my name or who mentored me, I had reached the glass ceiling. The best I could hope for was to become the next regional vice president: barely cracking the six-figure income mark, chasing people for their sales numbers and expense reports, monitoring employee's calendars for sales meetings, and kissing arrogant high performer's asses because they, single-handedly, fulfilled the monthly sales target. It was crystal clear to me that I didn't want to be what I had worked so hard to become.

Freedom Seeker | 13

One morning, after yet another sleepless night, I was running cold water over my face to reduce the swelling under my eyes when I promptly decided to take a little extra time before heading into the office. It was going to be a long day. First, eight hours of running life insurance quotes, replying to emails and dreaded sales calls, followed by an evening of schmoozing clients at an impressive appreciation dinner cruise through English Bay sponsored by one of our primary insurance company partners. When I arrived merely thirty minutes later than my usual eight thirty start, I met my boss waiting for me, leaning in the doorframe to my tiny office, one hand on her hip and her foot tapping the carpet impatiently.

"Did we sleep in this morning?" she asked provocatively.

"No, I needed a little extra time considering the fifteen-hour day ahead of us," I replied meekly.

She rolled her eyes and snarled, "That's not how we do things here in Vancouver. Don't let it happen again."

My jaw tightened and my ears began to burn. I squeezed out a nod of acknowledgment and turned to my computer. When I knew she was out of sight, I smacked my balled-up fist on the desk and took several deep breaths to calm down. I was furious that she chose to berate me in front of my teammates for something so inconsequential just to make an example of me. Now I was even more stressed about going to spend my evening pretentiously caring about people I only knew professionally and barely at best. It was exhausting for me to relax my social anxiety and have a good time, all the while managing to not overindulge on the free booze.

Later that night, five hours after we boarded the Star of Vancouver yacht, I was ready for bed. The VP of the sponsoring company hailed a cap for us girls, and as he thanked his team once again for the hard work and late night they put in to impress our clients he said,

"Now, I don't want to see you in the office until tomorrow afternoon. Sleep in and take some extra time!"

14 | Yvonne Winkler

That did it. All my discontent and frustrations with the patriarchal system and the subjugation of the corporate world, the built-up bitterness for the unappreciation my boss, a woman herself, had shown me, erupted. I bolted out of that cab and spent the rest of the night formulating a plan. I desperately needed a taste of real freedom.

As the days grew shorter and the clouds thicker I began to spend more and more time in the public library. I still needed to write and pass the CFP certification exam if I wanted to get my money refunded through our corporate continuing education program. But I already knew that the designation wasn't going to buy me the happiness I was craving. In fact, all it would do is keep me on this soul sucking hamster wheel existence, talking about things that didn't light me up with people I couldn't relate to. During my study breaks I wandered into the personal development section of the old library where I found Phil Keoghan's *No Opportunity Wasted: 8 Ways to Create a List for the Life You Want.* I was hooked within just a few pages. Phil's account of his near-death experience and how he overcame fear inspired me into action to seek my very own Amazing Race.

Two years to the date when I moved to Vancouver, I walked into my boss' office with a dry mouth, flushed cheeks, and a tightly gripped letter of resignation. I had rehearsed the same line over and over that I could use no matter what my boss's reaction was. "I'm sorry you feel that way. This is something I need to do for me." Determined not to be persuaded out of my decision, and prepared that they may walk me out of the office immediately, I was ready to cash out my savings, trade my heels for a pair of Merrells, and purchase a one-way ticket to Europe.

A month after I resigned, my friend Siobhan hosted a big farewell party for me complete with personal chef specials and vodka martinis. I felt loved by this group of mismatched singles who had formed a bond over weekend brunches and long walks on the beach. *Why am I leaving this?* I wondered looking around the room filled with laughter, shenanigans, and music. I had finally found my posse and a place I belonged. Since that night of deciding that I wasn't going to slave away under this false freedom anymore and I let go of striving, achieving, and competing, I had found a quiet happiness. There were moments I felt so intensely happy walking through a bustling Granville Island market, hearing the seagulls cry and taking in the scent of freshly baked bread, it would bring tears to my eyes.

The next day my dad loaded what few personal items I had left into his F150 pickup as tears welled up in my eyes and the muscles tightened in my throat. He was always the man power I called when I needed a hand with moving and this time was no different. He drove from Calgary to Vancouver to bring me "home" from where my adventure would begin.

What was I doing leaving all this now? But I couldn't go back. I couldn't undo my resignation from my job. I had personally trained my replacement. I would look like a flake, a loser crawling back to my comfort zone. No, I had to see this through and follow my heart that was aching for more.

My parents' doorbell rang, pulling me from my thoughts of the past. My friend Jennifer stopped by to give me one last hug and a gorgeous, brown leather-bound journal with linen paper to record all my experiences. Jennifer came into my life like a breath of fresh air, replacing me from my marketing assistant role in Calgary. Bright-eyed and quick with the sarcasm, she had me in stiches as she demonstrated the dance version of the in-flight safety instructions the attendants give before takeoff. When Jennifer joined our team we quickly got to know each other as I became her trainer. It was like riding a bike. All those hours in my childhood bedroom playing teacher paid off.

Over a common desire to kick cancer's ass one step at time at Tom Baker's Walk to End Breast Cancer, we became besties outside of work as well. Jen had a definite plan of where she wanted to be by the time she turned thirty; wife, mother, white picket fence.

We attended each other's birthday soirees and Christmas parties, and consoled each other's heartbreaks over bottomless bottles of red wine.

When I moved to Vancouver our friendship broke into a natural hiatus. We still called each other occasionally, but we filled our friendship buckets with new people who we could go and have a beverage with. When I announced my wild globe-trotting plans to her and a few others at our annual candy cane Christmas party, she seemed a little sceptical. She knew firsthand how difficult it was to get an invitation to the boy's club. She also knew there's no point trying to get between me and what I set my sights on.

Now that the day was here for us to part ways for an undefined period, and more than a hop, skip and jump away, I watched her gulp down her pain and put on her best happy-for-you face.

"I'm really jealous," she said releasing me from a hug. "You'll soon get to hang out by the Aegean Sea, and sip wine in Rome while I'm stuck here kissing ass to pay my bills. You're so brave to just walk away and find a new adventure."

I smiled, needing that boost of courage.

2
BREAKING FREE

East Germany, September 1989

We assembled as we always did, at the beginning of the new school year in the gym of the Arthur Geisler Oberschule, Penig. Marching in, and I mean marching, in order of class ranking, first grade one, two, three then four. I was starting grade five that year which meant I was no longer a young pioneer but a Thälmann pioneer, named after the former leader of the communist party, Ernst Thälmann, and distinguished by graduating from a blue triangular necktie to a red one.

On special occasions such as national holidays and the first day of school, we were all required to wear our uniform comprised of a white, collared shirt, necktie, and blue skirt or pants. Everybody knew that but there were always one or two kids who stood out in the perfect square formation wearing plain clothes. My stomach dropped a little for them as, before the assembly was over, they would be called, one by one, to stand in front of the podium facing five hundred of their fellow students while our principal made an example of them for their negligence. The last to march into the gym were the grade tens, the graduates for that year called the *Freie Deutsche Jugend (FDJ)* (Free German Youth).

Our principal, a giant, purple faced man, stepped to the podium. "Welcome students and staff," he sternly said into the microphone. "As we begin a new school year, let's begin by remembering what unites us. *Für Frieden und Socialismus, seid bereit.*" (For peace and socialism, be ready.) Every student and staff member in attendance stiffened their right hand, placed it over the middle part of our head with the thumb lightly touching the scalp and the little finger facing skyward and shouted in response, "*Immer bereit!*" which meant always ready.

Assemblies like these stressed me out to a high degree. I was so scared of being admonished in front of my peers for not having the right answers or disobeying the rules that I regularly threw up in the classroom sink — much to the disgust of my fellow classmates. I'd lay out my clothes the night before so I couldn't possibly forget to wear my uniform. Barely able to sleep that night, my stomach churned so badly that I was never certain which end I would lose breakfast from first, so I just didn't eat. I fought a hot tingling that was creeping up my spine, usually a sign to find a bathroom and fast, by quietly whispering to myself, *you're OK, you're OK.* After all, I couldn't break formation, that would definitely draw attention to me. I softly tugged on my necktie, check. Glanced down, white shirt, blue skirt, check. *You're OK.*

Starting that school year was particularly nerve-racking for me as I knew a dangerous secret that could get my parents into serious, life-and-death trouble, a burden I found impossible to bare as an eleven-year-old. I usually couldn't wait to see my friends at the bus stop and tell them about all the new adventures we had on our family's annual summer camping trip. That year was different.

While the principal continued his shouting to the assembly, my thoughts drifted back to a summer morning on a little campground in Kecskemet, Hungary. It had started out like any other day. I woke up to the smell of the little propane burner working hard to bring the titanium kettle to a boil. Mom setting the folding table with our blue, red, and green plastic cups and plates. Dad was reading the paper, which I thought was weird because he didn't speak Hungarian. Folding it back to a quarter of its size, I noticed the front-page picture of border guards and barbed wired fences.

Freedom Seeker | 19

"Today, we'll go for a drive and explore," he said with a decided tone.

After breakfast we zipped up our tent and crammed into our two door, crème-colored Trabant, a micro version of a car manufactured by and for East Germany, with a mustard yellow roof, and drove into what I deemed boring countryside. Nothing but fields as far as the eye could see. Suddenly, out of nowhere, a car appeared behind us, and before I could decipher what was happening Dad pulled over, looked over his right shoulder and directly at me.

"Stay down and don't look!" My mom and dad exited the car.

Telling a kid not to do something was futile at the best of times and sensing the imminent danger, I peeked through the back window long enough to see my mom and dad kneeling in a ditch with two men in uniform pointing guns at both of their heads. *Who were those people? What had we done? What was happening?* I heard another vehicle pull up and lifted my head again enough to see a tall, half balding man in a trench coat approach my parents with two more soldiers carrying machine guns. I quickly ducked back down and began to cry. I couldn't make out anything they were saying. Confused and terrified for our lives, I was trying to put the pieces together but my eleven-year-old mind couldn't figure out what rule we had broken to get in this kind of trouble.

We weren't religious people because communism and faith do not play well together. Karl Marx said so. And yet, by the grace of a higher power, they finally released my parents from this interrogation at gunpoint in the middle of a ditch somewhere in Hungary and escorted our car until we were out of the "zone". I'm not sure how I knew all the ins and outs of government surveillance and espial, but I knew it was then safe to ask:

"What zone?"

Mom and Dad looked at each other and staying true to the golden rule our family had about telling the truth to maintain mutual trust, Dad began to explain that we were on this camping trip with the intention to flee East Germany. He had been following the news and learned that Hungary and Austria relaxed their border patrol, increasing our odds of surviving escape considerably from the original plan, he assured me.

"What was the original plan?" I asked still in shock from all that just happened.

His eyes, filled with dread, met mine in the rearview mirror. "We were going to swim," he whispered.

I had no idea that all those swimming lessons were in preparation for an insanely high-stake escape attempt via the Danube, the second largest river in Europe, and through former Yugoslavia.

"Attention!" The loud order from the purple faced giant brought me back to the assembly. In unison, we turned ninety degrees to the left and marched out of the gymnasium in reverse order.

I took my seat at the narrow two person desk on the far right side of the room, closest to the door, just in case I had to make quick exit for the bathroom. Our new secondary teacher, Frau Frauenheim, introduced herself and while she laid out the curriculum for the year, I wondered what it would it be like to shop in stores that had real jeans and fashionable, colorful clothes instead of wearing Mom's homemade designs with an iron on Puma sticker.

I couldn't imagine what life might be like on the other side. To eat oranges more than once a year or a banana without brown spots. To watch a movie that wasn't censored by the German Democratic Republic or read a book with different opinions or viewpoints than the socialist ideology. Oh, and to travel to Spain, Rome or London. To have as many Toffifee as I could fit into my mouth and to buy more when the box was empty. The thought of having access to any of this was a dream. A dream my family was willing to risk everything for. My head was spinning out of control, as was my life. *What will happen to us now?* I wondered?

Immediately upon our return from Hungary my parents began the process of legally obtaining an exit visa. We were drilled on communist ideologies, Marxism and the rules that governed the GDR from a young age. I knew that our attempt to turn our back on the GDR, especially without their approval, was deemed an illegal border crossing. Now that they had submitted a formal application, the threat was worse. We could face lengthy and severe harassment, criminal prosecution, and imprisonment. If they wanted us gone our disappearance would be swift and clean, nobody would ever be the wiser.

We were quickly approaching the 40th anniversary of the German Democratic Republic which would be commemorated with a military parade and torchlight procession through the center of our capital, Berlin. Students from across the country were selected for this honor, carrying the Free German Youth torches, and my friend Steffi had been chosen to represent our school. Although I was always considered a dedicated student, I was not among the chosen ones. One of many consequences for our now publicly stated intention to leave. Mom was unable to find work; nobody in the region was willing to risk hiring her for fear of losing the few privileges they had. A clever multipronged technique enforced by the Ministry for State Security, also known as STASI, to shame us for the betrayal we'd committed to the republic and, perhaps more importantly, to label us "unmannerliness enemies of the state", thus making us undesirable immigrants to the West.

An unmistakable feeling of uncertainty and fear was around us at every corner. Mom and Dad were barely recognizable with deep creases from worry on their faces and nothing more than skin and bones for the lack of appetite the ongoing stress was causing. It was devastating to watch my father so on edge. He was the only man in my life I could depend on and trust without a doubt. I suddenly had a strict six o'clock curfew, which I disobeyed only once. When I came home and tried to pull my usual "Oh, my watch

22 | Yvonne Winkler

stopped," excuse, my dad was so scared that the STASI had taken me that he blacked out with rage and Mom had to pull him off me.

My life was flipped upside down, debilitating apprehension about who we could trust, who would sell us out for a small privilege like a telephone line, and when they would come and take us away. *Was it really worth it?* I asked myself often as I followed the flicker of the candle in my window. I thought it would be incredible to see what the land of milk and honey was like, but I couldn't conceptualize having to say goodbye to my Omas and friends forever. That was the cost of freedom and the price of admission when you turned your back to the GDR.

About a month into grade five I was called into the principal's office, my cheeks flushed red hot, my heart was beating so fast that I got dizzy as I hesitantly made my way up the four flights of stairs to the teacher's lounge. One of my friends from 5a, Kerstin, had been called into the principal's office just last week. I had heard rumors that her mom had also applied for an exit visa, and I never saw her in school again. *Were they going to hold me hostage against my parents now?* I panicked as I knocked on the door.

"Come in," a woman's voice said. I pushed the door open timidly and just enough for my little body to slide through. "Sit there," the expressionless woman motioned to me, taking off her glasses and letting them fall onto a beaded chain around her neck. She got up from behind a heavy wooden desk consumed by a large typewriter and walked over to the closed principal's office door.

"Yvonne Winkler is here, as per your request."

"Bring her in."

Just hearing his voice made me quiver. For five years, I had dodged any reason to ever be alone in a room with this man. To my surprise, he wasn't alone. My former teacher, Frau Havel and my new teacher and Frau Frauenheim sat, waiting at a small round table. I sat in the empty chair next to the principal with my hands folded in front of my body, and stared straight ahead at a picture of

Erich Honecker, the leader of the German Democratic Republic. His picture was in every room of any official office and classroom. Just so we wouldn't forget who we bowed to. Choosing their words carefully so as to not frighten me and instead unlock a girl's hidden secrets, the two women began to inquire about the months leading up to our trip to Hungary. Terrified in the man's presence, I answered their questions truthfully. They were looking for incriminating evidence to find my parents guilty of attempted escape. A futile effort because Mom and Dad had not shared anything with me to prevent exactly this from happening, so they had to let me go.

Meanwhile at home, we sat on crates and ate from plastic plates. We had packed up everything we owned and either stored it in my grandma's attic or shipped it to my mom's girlfriend in West Germany, in anticipation that we would follow soon. Each night, Mom lit a candle and placed it in the window facing the street as a beacon of support to those who fled or who were awaiting their fate at the refugee camps.

On the morning of November 9th, 1989, by way of a hand delivered envelope, we received word that we must leave immediately. Our exit visa had been granted. My dad ordered us all to get in the car, "Hurry, before they change their mind!" he shrieked.

I hugged Oma Lisa and Oma Emmie who quickly rushed over to our house to say goodbye and to plead with my parents one last time to not leave. As we drove through our little village of five hundred people for the last time, neighbors and friends waved to me in the back seat, as I sobbed inconsolably.

After eight hours of inching our way to the border that separated us from the other side of Germany, we were welcomed by my mom's friend and her family. Margit was my uncle Harry's former romantic partner whom Mom had met while visiting Nürnberg three years earlier, for his fiftieth birthday, an occasion for which our government granted special travel permissions for

24 | Yvonne Winkler

"low risk", close relatives to go on a once-in-a-lifetime trip to West Germany. Uncle Harry left East Germany just nights before construction of the wall began in August 1961, and he was my favorite uncle because whenever he came to visit Grandma would have all her children and grandchildren over, stacked into the living room of her three-bedroom worker's housing cooperative apartment, for afternoon coffee followed by family dinner, more drinks and loads of laughter. It was pretty much the only time I saw my cousins on my mom's side.

Mom and Margit continued to stay in touch and grew a friendship that prevailed the breakup with my uncle. Margit became the kind aunt from the West who, on birthdays and Christmas, sent us packages with things we couldn't buy in the GDR; Toffifee, Jacobs Coffee, and a box of Frosted Cornflakes for me. She lived with her parents, Robert and Erika, in a beautiful timber framed three-bedroom house in Schwaig, a suburb of Nürnberg, in middle Franconia.

They had made room for us until we found an apartment of our own by disassembling their dining room and giving us the upper floor living space Margit normally occupied. It was a squeeze for the three of us sharing one small room, but it was temporary and we were grateful we didn't have to live in one of the refugee camps. All immigrants from the GDR were required to report to a nearby receiving camp where East German border crossers would receive a new passport and relocation assistance.

Because we had an address to give to the officials, my parents left Margit's house to get registered and collect our "welcome money" of fifty deutsche marks without me, so sparing me what Mom later described as the uncomfortable scene of thousands of families sitting on fold out beds two feet apart awaiting their fate. Instead, Margit thought it would be good for me to come along on her weekly trip to the supermarket.

As the touchless doors opened into the Aldi Store I was met by a pyramid of oranges. Next to it ripe, yellow bananas and delicious red apples. The produce section alone was the size of the entire little market we shopped at all my life. It smelled a lot better too. The unmistakable scent of quality coffee beans and chocolate hung in the air. When we passed the magazine aisle Margit said, "Why don't you pick one to read for when we get home?"

I had never seen so many magazines or chocolate choices. I stuck with what was familiar to me from the West packages Margit had sent us. One Bravo magazine, Hanuta wafers, a pack of Duplo Sticks, and one box of Kellogg's Frosted Flakes.

"They're grrrreaaat!" I whispered and then my brain went into shutdown.

I didn't speak for a week.

I grew up being told this escape was a crime that was punishable, usually by death. The trauma from the events in Hungary until the historical day in November, combined with the shock of seeing overstocked shelves with anything a child could ever imagine, had left me literally speechless.

3
LINE IN THE SAND

Germany, April 2010

Tired and sweaty I dragged my suitcase with both arms up the tiny metal stairs onto the train that was leaving Frankfurt station in less than a minute, for Leipzig. A busy mind had kept me up all night replaying scenarios about what the next year would look like and rehearsing lines like an actress about what I'd say to the people I left behind two decades ago. Over the years I had gone back for Grandma's birthday, a family reunion or momentous anniversary, but those were short overnight stays with little time or desire to walk down memory lane. I had built a new life with new friends and the distance between us grew bigger, figuratively and literally, with each year that passed. Since we moved to Canada, I hadn't seen anyone in at least five years.

Will my aunt even recognize me? I wondered as the trees and houses whizzed passed. I spent months preparing for this *back to my roots and beyond* trip as I fittingly named it. Living in Vancouver as a single thirty-something provided me with ample practice doing life solo. Starting with small steps like going to a movie alone or entering a restaurant and confidently requesting a table for one. It was awkward at first and I hated the "Just for yourself?" question every hostess asked with raised eyebrows. But as my confidence grew, the feeling of empowerment became irresistible. I soon found myself venturing out beyond the downtown grid and exploring the islands around Vancouver on weekend escapades. I loved being the captain of my own ship, the steward of my adventures, with little to no consideration for anyone else's likes and dislikes. It was ecstasy.

I stepped off the train onto the marbled platform and was instantly reminded how pristine Germany was. *"Hallo meine liebe Yvonne!"* My aunt Marlies moved hastily toward me waving her hand. Marlies was the youngest of three with two brothers. A kind-hearted, spirited woman with a sense for adventure but always worried about something. After the fall of the wall and months of sorting out property rights in the rubble of forty years of communism, Marlies and her husband Bernd purchased the house my parents had built, at my grandma's request to keep it in the family. Their daughter, Nicole, was only three then and quasi grew up in my footsteps. Her bedroom was where I once played, studied and put myself to sleep listening to cassette tapes of "Hansel and Gretel", "Frau Holle", and "Rumpelstiltskin".

I have five cousins on my mom's side of the family and two on my dad's. Nicole and I had always been the closest, despite our seven year age difference, mainly due to their efforts to stay involved with us when we left this little town, Tauscha, East Germany. It was serendipitous to be able to begin my trip in my childhood home.

As we drove down the winding road barely wide enough to fit a Volkswagen Golf, the first thing I noticed was how much smaller everything appeared than how I remembered. Inside the house, the hallway from the kitchen to my old bedroom once served as my practice ground for learning to leapfrog. Now I could reach the door in merely a few strides. As I stood in the middle of my old bedroom, I could still see where the oak framed twin bed used to sit against the right side of the wall, with a matching nightstand and my red Sony cassette player. A three door wardrobe next to that. Dad constructed a custom desk for me out of an old cabinet. It was the coolest thing complete with built-in shelves, light, and a space-saving drop-down tabletop that I could lock. Perfect for an eleven-

28 | Yvonne Winkler

going on twenty-year-old. He was creative like that, and we had to be. We couldn't thumb through catalogues and order the latest IDÅSEN from Ikea. I had begged my mom for months to paint over the childish, giant Goofy cartoon on my wall until she finally wallpapered the room with a simple, textured pattern and completed my grown-up space with a large, round mirror. It was the envy of all my girlfriends.

I seemed to always be in such a hurry to be treated like an adult. At least, I didn't want to be belittled with cute words of endearment. It's part of the German formality to call older relatives by their title as a form of respect. I refused to call my aunt "Aunt Marlies" because the grown-ups didn't. Perhaps my biggest feat of rebellion and proving my independence was to hold burning amber in the form of a narrow paper tube containing tobacco between my little fingers.

Die Bude was what we called our old hay shed. After Mom and Dad became flourishing pub owners, and most of their waking hours were dedicated to keeping all aspects of this small but mighty watering hole for the community gardeners running, my friends Claudia, Silke and I turned it into our hideout. It was perfect because it was close to home, yet far enough away from the careful watch of an adult.

My desire to smoke didn't have anything to do with Mom and Dad working in a pub, where the regulars puffed away for hours while catching up over warm beer. I sharpened my pencils to the length of a cigarette and pretend smoked, like I had seen all my relatives do, long before that. Maybe my six-year-old brain worked out that smoking was the way to get closer to my grandpa. Opa Herbert was my only living grandpa, and not exactly the warm and affectionate figure I heard about in my bedtime stories. His harrowing experiences of being in the marine corps, followed by five years postwar slaving in a coal mine in a Russian concentration camp, left him mostly distant and broken. I used to sit, with my feet dangling over the edge of the brown sofa in their kitchen, quietly observing him as he opened the silver cigarette case, his right foot on the wooden stool by the window, and as he blew out the smoke he leaned his body forward onto his leg as if he was starboard looking for new land.

The pub provided me with easy access to cigarettes and the privacy we needed to smoke them. One day I mustered up the courage to deposit the equivalent of three dollars into the bar till and snuck a pack out the door in my satchel. Claudia and Silke met me in *der Bude* where we sat on a makeshift bench I made from an old pig trough and a two by six stud board. We watched with amazement as Silke showed us up by blowing smoke out of her nose. With a flick of the spark wheel and a deep inhale, I brought in too much smoke and began coughing so hard I nearly threw up. Determined to not look like amateurs we practiced smoking properly in our hideout whenever we could.

I loved everything about cigarettes, except the taste. The thin, smooth paper perfectly rolled. The sweet scent of the tobacco, with a faint hint of earth and dried grass. The way a fresh pack excited me, and most of all, how cocooned, cool, and mature I felt holding one between my index and middle finger.

It was a rainy Saturday afternoon when Mom jolted into my room, threw the newly opened package on my desk and hissed, "Next time you want to kill yourself, at least buy the cheaper brand."

Mortified, I stared at the paper package in front of me. *How did she find out?* "I'm sorry," I whispered through the big knot in my throat. But Mom had no time for my excuses. That was the first time she'd grounded me, and I didn't pick up another cigarette for three years.

"Do you still have the old hay shed?" I inquired turning my attention back to my aunt who was now tending to the still frangible plants that would soon bedazzle the wrap around balcony all summer long. Her and uncle were avid gardeners and they had done an incredible job picking up where Mom and Dad had left off.

"No, we took all of that and the rabbit hutches down," she looked up. "Do you still smoke?" she asked knowingly.

"Nah, I finally quit that stinking habit. It's been four years and counting," I said with pride.

"Ach, I wish Nicole would quit too. Bernd still has the occasional drag. Helps him relax at the end of the day. It's such a gross habit." She sounded worried and defeated. "I only hope you don't start again while you're here".

"Me too," I replied. It hadn't really occurred to me that after all this time I might be tempted by old ways. Smoking had been my crutch for fourteen years and it was one of the hardest things to let go. It was my comfort when I was feeling nostalgic, it had been my gateway to making new friends in the places we moved to. It was the one constant I could identify with and how I belonged. Later, in the corporate world, it was the bridge between me and the administration girls who every hour made their way downstairs for a quick break. I eventually quit because it didn't fit with the vision of who I was becoming; successful, fit, elegant, stoic, and a respected business leader.

Canada, unlike Europe, had made huge strides towards trimming healthcare costs and encouraging a clean, non-smoking environment. In Calgary and Vancouver restaurants and bars hadn't allowed indoor smoking for years, making it socially unacceptable and banishing the dwindling group of smokers to enjoy their cigarette in frigid temperatures. The smell of smoke that clung to my clothes was stenchier than when we were all nose blind from it being everywhere. I was beginning to feel increasingly uncomfortable and hyper aware of my failings with this unladylike habit. It had to go.

"So where are you going first?" My aunt opened the fridge and grabbed a carton of milk. We were baking Oma Emmie's favorite, apple strudel, for the same afternoon when the rest of the family joined us on the balcony to celebrate Easter Sunday.

Freedom Seeker | 31

"I've booked a flight to Barcelona. After a few days there, I'll take the bus down the coast to Malaga. Hopefully, I'll find work somewhere along the way," I said in between licking the spoon clean of the last bits of dough.

"Oh Yvonne! Are you sure you want to do that?" My aunt furrowed her eyebrows, full of concern. "Barcelona is so dangerous. Many tourists get mugged there every year. And you, you're all by yourself."

Here we go again, I thought as I rolled my eyes and turned away so she couldn't see any uncertainty in my face. "I'll be fine."

I had been through this dialogue many times since the announcement of my plans to backpack through Europe on my own, and I had weighed the pros and cons carefully. I wasn't reckless and I was very aware that I had never traveled anywhere alone where I didn't speak the language. I also had to figure out transportation and accommodations on a shoe-string budget but I was determined not to let fear stand in the way of my freedom anymore. Many women much younger than me travelled alone all the time and they were fine.

"I'll keep my wits about me, and I won't venture out at night, OK?" I added.

My mind was made up. Nothing was going to stop me from following my yearning to be free, like a child from the safe grip of her mother's hand; through the barriers that corporate, social convention, and well-meaning friends and family wanted to impose on me. Ever since November 1989, when we left everything behind for a chance at an unrestricted life, my family became freedom seekers. Along the way, I stepped into a few traps but my quest for autonomy over my life was unleashed once more that day my boss gave me a tongue lashing for daring to color outside the lines.

Having lived through both systems, I could see the similarities between the communist regime and capitalism. What were physical barbed wire fences and political brainwashing to control the folk of East Germany, were now the shackles of

32 | Yvonne Winkler

overwhelming debt, consumerism, and the grand illusion of certainty. For example, I thought that my paycheck was certain and I became very compromised by it. My health, happiness, freedom, and my potential were all compromised in the name of certainty. And then one morning I arrived at the office, and I couldn't take it anymore.

From early on, I had been told how I had to feel, who I needed to be, and what I can and can't do. To uphold the communist mindset we were discouraged to think for ourselves, dream big or desire a different life. The survival of the regime depended on conformity and oppression. Our aspirations were as caged as the people who lived behind the wall.

Everything I did, from smoking to working endless hours of overtime, was driven by the desire to feel a certain way. And because I didn't know how to read my feelings, and instead learned how to suppress them, I relied on measurable certainties. But here I was, finally rid of the shackles. I had no children, no mortgage, no employer and no bills. I was no longer restricted by a government. I could choose a different path. I had no idea what I was leaping into and how it would end but I knew I needed to get away from anything and anyone tying me down. My parents had risked everything without knowing what would happen. It was the soul's quest for freedom that kept us alive and got us through the iron curtain.

"Freedom isn't anchored in certainty," my coach once said to me. "It's your willingness to take risks and leap that will move you forward."

I needed to explore for myself that I could not only survive but thrive on my own. I was tired of reaching for the stars and coming up short, or a wagging finger pointing out that I'm not strong, pretty, intelligent or something... enough.

***

Spain, 2010

I woke up to the sweet, berry-like fragrance of fuchsia bougainvillea vines in front of the window and a fresh sea breeze caressing the soft purple curtains. A familiar sound of seagulls fighting over scraps and a whisper from the ocean, like it was calling my name. I swung my legs out of the lower bunk bed and rubbed my eyes. *This might be it,* I thought. I immediately loved the bohemian vibe of this place. A stirring in the kitchen just across the hall indicated that someone, probably Rafi the hostel keeper, was up preparing the usual backpackers' continental breakfast consisting of toast, strawberry jam, tea, and coffee for the guests of The Melting Pot, in Tarifa, Spain. This was my fifth hostel stay and I had quickly learned to take full advantage of this simple but free meal. Rising earlier than everyone else was a new thing for me but made sense since I was usually in bed early too. I had kept my promise that I'd made to my aunt to stay alert and hadn't had any alcohol since I left the familiar surroundings of my former hometown in Germany. Being highly introverted, sitting around with strangers who were partying without the nerve calming effects a drink was awkward. Instead, I would take myself and my pearl white Samsung Notebook to bed and watch whatever free episodes of *How I Met Your Mother* I could find. It kept me connected to a world I knew; a small comfort when everything else was unfamiliar.

I had arrived in Tarifa midafternoon the day before via bus from Malaga, where I had taken the advice from a woman about my age in geo print carrot pants and Rasta locks, that my chances of finding what I was looking for were better here. I believe she told me that because there were only so many jobs in Malaga and there were lots of us. Leaving Malaga behind was easy. I didn't really get the sense that the coastal town had what I was looking for. Sure, it checked the box that it was a place on the ocean but tourism had stolen its charm. The dark sand beaches were crowded even in the early part of summer. I could only imagine the sardine-

34 | Yvonne Winkler

like stack once July rolled around. The shorefronts were lined with buzzing beach bars and watersport rentals. It was more cosmopolitan than the visions in my dream.

I grabbed my towel, toothbrush, and Elizabeth Gilbert's latest book *Eat Pray Love* and left the other sleeping travelers as quietly as possible.

"*¡Guapa! ¡Buenos dias!*" Rafi greeted me. He towered over me, and I figured he was in his late thirties but it was hard to tell with sun-aged skin, windblown ashen locks and the grey arc over his right cornea. He invited me over to grab a coffee and used the opportunity to give me a quick tour and instructions for shared kitchen space etiquette. Never having been one for early morning chatter, I smiled and nodded as he explained where I could store my groceries and then I excused myself to the sun-bleached rooftop patio overlooking the Strait of Gibraltar.

Coffee and book in hand, I parked my butt on one of the wide daybeds, constructed out of old pallets and a well-used foam mattress, and flipped to the earmarked page where Liz explained to her best friend Delia that she has no pulse anymore and desperately needs a change from her life. It felt like we were on a similar quest, and while she ran off to Italy, I found myself looking for a new pulse in Spain. It had been an aspiration of mine to add a third language, Spanish, to my repertoire and I knew from my previous experience learning English that the fastest way to become fluent was by being immersed in its culture. I hadn't had many opportunities to practice Spanish in the two weeks I had been in Spain. Most of the hostels in Barcelona, Valencia, and Malaga were occupied by young people speaking or practicing English. *Maybe once I find work I can enroll in some Spanish classes*, I thought.

"*¿Qué tal?*" Rafi interrupted my thoughts, joining me after the breakfast rush. "So, what brings you to Tarifa?"

"I'm looking for a spot to hang out for the summer and pick up a few shifts. Do you need some help here in the hostel? I can clean rooms and make breakfast," I said with a bit more energy now that I was fully awake.

"We've got that covered here," Rafi said, "but the owner, Louis, will be in town tonight. I'll let him know you want to meet him. He might have something for you in Cádiz."

I was beginning to see that it wasn't easy to find temporary work in Spain. The economic situation in most parts of the country wasn't exactly flourishing and everyone needed to do what they could to survive. Locals found themselves constantly competing with young globetrotters like me who were looking to subsidise their adventures and were willing to work for far less than what the people of Tarifa needed to feed their families.

When we moved to Canada I had made the decision to keep my options open, be a permanent resident and not give up my German passport. That choice proved very practical as employers viewed me as a member of the EU. Unfortunately, because of the political and economic climate and a hostile attitude to some countries in the EU, particularly Germany for turning off the bailout tap, I was careful not to flash around my red passport too much.

"OK, I'll make sure to connect with Louis tonight then, thank you."

If I could land a shift in exchange for room and board it meant I could afford to spend a few months in Spain and live out my beach life fantasy complete with sun-kissed skin, Bob Marley beats carried in a summer breeze, flowy dresses, and sunset parties.

Satisfied with my efforts on the job search, I loaded up my day pack with water and snacks, rolled up a towel, and strolled down to the beach. I had heard of a must-see spot called the giant dunes of Punta Paloma Beach, roughly a thirteen-kilometer trek along the shore where the Atlantic Ocean and Mediterranean Sea meet. Beachcombing was one of my favorite ways to pass time and my soul was aching for some alone time. Sharing tight spaces, including showers and bathrooms didn't leave much room for

36 | Yvonne Winkler

privacy and that was beginning to get to me. My body was stiff from being squashed on trains and busses and I needed a good stretch and exercise. Without further thought, and the dunes as my target, I took off my sandals, clipped them tightly to my pack and began to walk the soft, golden sand, leaving the white, multifoil arched buildings of Tarifa behind me.

Tarifa is an ancient fishing village with a violent history of war and conquests by the Moors, the Romans, and the Christians. It had castellated walls and a statue of Jesus Christ so tall it could be seen from the coast of Africa, nine miles across the Strait. The statue stood as a reminder to when the Spanish catholic kings drove back the Moors for the last time.

I walked along this wide-open stretch of sand with my sights on reaching the infamous giant dune, and I quickly fell into the rhythm of my surroundings. It was calm, and the waves gently deposited shells I had never seen before at my feet. It made sense that different types of mollusks ate different diets and would leave behind a variety of seashells. These ones looked like angel wings dipped in chocolate. I picked them up one by one, only keeping the unbroken ones, zoning into the magical sound of the waves as they raced up the sand and then receded back into the comfort of the big blue.

Flooded with emotions, I thought about my final weeks in Vancouver. I recalled my boss's words as I handed her my resignation. "After everything we've done for you," she had said in a slow and disappointed tone. Having my motivations or moral ethics questioned was triggering for me. I didn't think I owed them anything beyond all that I had already given; those words cut me deeply and brought on an all-consuming rage. I carefully climbed over a spectacular display of rock formations that resembled the spikey spine or razor-sharp teeth of a dragon. I wiped the tears from my eyes to see where I stepped. *Don't need to twist an ankle or gash my knee on these,* I thought.

Freedom Seeker | 37

When I was passed this obstacle course, I stood there for a minute wondering how these rocks would have come about. So sharp they couldn't have been there long. The wind had picked up, a welcome breeze as the sun was climbing to midday. I was alone for as far as the eye could see, not a single soul around, and I was perfectly OK with that. I was so engulfed in all the feelings that arose, I wasn't scared at all. If anything, it was probably the first time ever in my life that I occupied that much space by myself and I could let down my guard and be me.

The shell of the identity I had carefully crafted each morning before heading to work was cracking. Every footprint in the sand was another story that got rinsed away with the next swell. In this almost hypnotic space I reached parts of my inner self I hadn't been to before. I laughed out loud, I cried softly and then bitterly, I kicked the sand and screamed.

After I had emptied out all the feelings, I suddenly saw colorful parachutes attached to tiny, toy figures in the distance. The wind was now blowing hard enough to wisp sand into my face, perfect conditions for wind sport fanatics. I passed an instructor who was showing a young couple the safety know-hows and procedures that were essential when you voluntarily harness yourself to a to a sixty-foot piece of fabric in twelve miles per hour winds.

Kitesurfing, the ultimate form of freedom, I thought. Yet, the fear of losing control to the elements and trusting my abilities was too strong and stopped me from signing up for the next class. I decided to stop here and watch for a while before going onto the final leg. I could see the dunes, but estimated I was still at least a thirty-minute walk away. Laying back on my towel, I stared at the cloudless sky allowing the sound of the waves crashing into the shore to take me back into reverie. I began to see what I was supposed to do. No more wasting my time in a job that didn't light me up for the sake of earning a number that would determine my social status. I began to envision a beautiful space where people come to rejuvenate and heal.

I would create a place that had saunas and plunge pools. Meditation tents and journaling space. It came fast and with crystal clarity because it wasn't an entirely new vision. I had drawn up the plans for a wellness spa before, it even made it onto paper. But now I knew that I wanted to make it happen for real. It would be a collective with different healers coming together, sharing the space and customers. I thought about learning an ancient Indian healing system called Ayurveda as part of my contribution to the temple. Healing the whole body through the stomach had my curiosity for several years in my efforts to find answers for the ongoing issues with my gut.

A calmness wrapped around me and the tickle of excitement in my body assured me this was my purpose. I sat there, the water creeping up to my toes and then receding as I looked upon Africa, and felt a little like the Andalusian shepherd boy, Santiago, in Paulo Coelho's masterpiece *The Alchemist*. Only I wouldn't venture into the Sahara Desert to find riches of gold. I decided that if I returned to Canada, I would never give my time and talent to an organization that cared more about profits than people. It was up to me to create the life and world that I wanted to live in. *I'll look into schools for Ayurveda in Canada and India and, I'll have to write a proper business plan for the Wellness Temple,* I promised myself.

I returned to the The Melting Pot in the late afternoon, hungry, exhausted, and emotionally drained from the walk. I went straight to the kitchen to retrieve a slice of white bread and aged ham and turned to the sitting room.

"*¡Guapa Yvonne! ¿Qué tal?*" Rafi was sitting with another man at a table having a beer.

"*Bueno,*" I smiled, my eyes fixed on the dewy glass of yellow liquid. It looked so refreshingly good but I remembered my promise not to drink. *Surely one couldn't hurt,* I speculated.

Freedom Seeker | 39

"Yvonne, this is Louis, the owner of the The Melting Pot," Rafi gestured to the man with buzzed black hair and a mustache.

"Nice to meet you, Louis. You have a real gem in Rafi here looking after your hostel which, by the way, is really nice," I charmed.

"Thank you. Can you speak Spanish?" Louis inquired.

"*Muy poco*," I said hesitantly.

"*Claro*, I need to practise English," Louis said unconcerned. Rafi explained to him in Spanish that I was looking to speak with him about possible work in his new hostel in Cádiz.

"*Sí*, you can cleaning rooms, OK? I need someone speak *español para la recepción*. We open in one week. You help with *preparativos finales*?"

I couldn't believe it. *Was it really that easy? Did I just land a job for the summer?*

"How do you pay? Room and food?" I tried to sound cool.

"No, I need beds in hostel for guests. I pay you in euro, OK?"

Even better, I thought. If I was getting paid, I could find an inexpensive room and budget for my own food.

"We have party here tonight. You come!" Louis got up and left.

"Well, that calls for a celebration," I grinned at Rafi who was already at the fridge opening a beer for me.

"Mark it to the room or pay cash?"

"Mark it please," I took a long swig of the chilled liquid and immediately felt the warmth of the booze rush through my veins.

The next morning, I pulled my eyelids open with my fingers. I had forgotten to take my contact lenses out, and that wasn't the only thing I had forgotten. I remembered meeting the new girls who arrived that afternoon; Katja from Germany and Natalie from the United States. I recalled assertively pushing away one guy who couldn't take 'no' for an answer. I had learned Spaniards are very stubborn and persistent. Beyond that, only the pictures on my point and shoot helped me piece the night together. Another thing I learned about the Spanish people is that their evening outing didn't start until eleven o'clock at night and that included dinner.

I lifted my head carefully off the pillow and assessed my condition from a seated position. A routine I learned some time ago when the drinks flowed, and the glass was seemingly bottomless. Sudden movements only caused dizziness, and if I didn't make it to the bathroom in time it would extremely embarrassing, especially since I just landed a job with the owner. *What kind of impression would that leave? A girl who can't handle her liquor.* I got up as agilely as I could and made my way passed the front desk and the new arrivals to the showers. I was leaving for Cádiz in a few hours.

***

With my head resting on the window, I closed my eyes to shield them from the brightness of the sun dancing on the ocean beside me. The bus ride to Cádiz took about two hours along the Costa del Sol, which was plenty of time to contemplate how I could let myself get so drunk. I was in a foreign place, didn't speak the language and no one had my back. *How could I be so careless?* I wondered. I could come down on myself harder than anybody. Not to mention that drinking and the subsequent shame led to a depression that could last a few days.

When I arrived in Cádiz I was in agony as I impatiently sat through the welcome and intro tour the hostel keeper rattled off. I remember vaguely hearing something about throwing toilet paper into the garbage can beside the toilet and under no circumstances into the toilet, and then I rushed to my assigned bunk and crashed hard.

When I woke up an hour later, a loud gurgle from my stomach reminded me that I hadn't eaten anything yet. I grabbed my wallet and set out to find a burger joint. It was the unquenchable thirst, a pounding head and inexorable sweating that made me wish

I hadn't drunk so much. At home I would have all the care supplies to cure the nasty effects of a hangover nearby. I had never been to Cádiz and had no idea which street to take. A greasy burger is all I could focus on. After aimlessly wondering around, I settled on a restaurant with an English menu, in the middle of Plaza de las Canastas. In my haste, or perhaps it was the hunger, I had forgotten my translation book. When the waiter came over with the menu, I asked if I could please see the English menu to which he just shook his head and mumbled something I didn't understand.

"No English, Spanish only," he said angrily.

I was a bit surprised. I had, perhaps somewhat naively, thought that they would be happy about visitors who would inject money into their hurting economy. Nearly on the verge of passing out from hunger I didn't have the energy to debate their huge sign at the entrance advertising an English menu. Instead, I pointed at something that looked like it could be meat and ordered *una* 'hair of the dog' *cerveza*. Forty-five minutes later, the waiter came back with a plate of five thickly sliced pieces of Spain's finest blood pudding. I gulped down the erupting contents of my stomach and said out loud, "Fuck it, when in Spain."

This wasn't the only time I would run into a distinctly brusque attitude from the locals. One beach bar outright refused to serve me a drink and asked me to leave. Being a foreigner wasn't new to me. I had faced discrimination due to my nationality before. *Suppose, I could have ordered in fluent Spanish, would that have made a difference,* I wondered? When I moved from East Germany to West Germany, I dealt with being the "Ossie". We spoke the same language, yet my dialect clearly identified the part of Germany I came from, which automatically branded me as less intelligent, poor, and inapt. I was excluded from high school history class in Nova Scotia because the teacher thought 'it would be best for me' while they covered the holocaust and Nazi Germany. The horrific actions of a single man have tainted German nationalism for seven decades. When do we stop punishing people for actions

42 | Yvonne Winkler

they had no involvement in? When do we stop hating people because they are different from us? I have experienced privilege and I have experienced discrimination due to my sex and nationality, unfortunately I learned more from the latter. I learned the Bavarian dialect so perfectly that nobody could tell where I was from just by listening to my speech. I mastered English fluently that many people were surprised to find out that it's not my first language. All that to not standout or be different. All in an effort to fit in.

I met Ana shortly after starting my job at The Melting Pot Cádiz. Ana was a funny, outgoing lady who leaned into Cádiz's university town environment by embracing it fully. She rented out one of her rooms and offered cooking lessons while teaching Spanish. Determined to learn Spanish fluently in my three-month stay, I saw Ana once a week before my shift at the hostel. I couldn't afford her room rate or lessons; however, we reached an agreement that she would teach me and I would help her around the house and walk her little dog Camarón, named after her favorite flamenco singer, Camarón de la Isla. Her laid back, hospitable, and kind nature made it easy to make her a friend and close ally. She introduced me to the vendors at the local farmers' market and showed me how to select the best scampi and clams for a one of a kind *paella de frutos del mar*. One evening Ana, her husband and I went to Plaza de San Antonio to listen to a local flamenco artist sing from her balcony into the cobbled stone plaza filled with a crowd of enthralled listeners. You could hear a pin drop as the singer stepped out, softly placed her arms on the stone balustrade and sang the first note, reminding me a bit of Madonna singing "Don't Cry For Me Argentina" only way more bona fide. It seemed Ana was determined to expose me to as much of her culture as she could, like she knew that it was the secret ingredient to mastering her language.

Cleaning at The Melting Pot was refreshingly mindless. I loved that it didn't require any mental energy, although it made up for it in physical endurance. I put my headphones in, turned on my ever-expanding 2010 playlist on my iPod shuffle, occasionally bellowing along, without any care for who might hear me or how off-key I was. It felt so nourishing to just be me. When my shift was done around one o'clock in the afternoon, I grabbed a *bocadillo* (a Spanish sandwich) at the corner of Calle de Sagasta and Calle de José Cubiles and ate it walking to the beach by my place.

As soon as I found a quiet spot, a few steps from the beach bar, I dropped my clothes and ran into the turquoise water to cool my body from the thirty degree humid air, followed by an hour-long siesta with the wind softly stroking my back. During my first week scrubbing floors and making beds, while tiny beads of sweat ran down my face faster than I could say "Holy hell it's hot", I completely understood the concept of siesta. It was absolutely necessary to replenish the body with lots of fluids and rest after working in the intense Mediterranean heat.

The room I had found in Santa Maria was about a twenty-minute walk from the hostel, outside of the city gates and right across from a beautiful stretch of white sand beach. The almost two hundred thirty square feet layout reminded me of my university dorm. Small, rectangular shaped with a single bed pressed against one wall, and a small desk, chair, and closet to the other. There was no window making ventilation impossible, which was probably illegal. But I spent most of my time at the hostel or at the beach and the price fitted my budget. Most nights, I went back to hang out on the rooftop balcony with the ever-changing guests of The Melting Pot. My friend Lucia or Kym, who worked with me, often joined me for tapas and drinks which, nine times out of ten, led to more drinks and moving the party to the many pubs or nightclubs in Cádiz.

One afternoon, about a month into my stay there, Kym and I finished our shift and went for a *clara*, a refreshing blend of beer and lemon soda, at a cafe I hadn't visited yet. I liked Kym. She was

44 | Yvonne Winkler

a free child. A modern-day hippy. She had dark shoulder length hair, very short, straight-cut bangs, and an undercut. Her ear loops stretched around dark ear skins and made her piercings look like big open holes. She had the most stunning big blue eyes outlined with perfectly thin shaped brows embellished with a side stud. Kym was from Belgium and spoke fluent English, Spanish, and French. She was an excellent person to help loosen my inhibitions and show me freedom.

"How do you like it here?" she asked lighting a cigarette and leaning back in her chair as she blew out the smoke.

"I love it! I'm so glad I've found work so I can spend the summer here and learn Spanish."

"How's that going?"

"Well, it's hard to take any group lessons as most are throughout the middle of the day when we're working," I said a bit disappointed. Having to choose between work, which enabled me to stay in Cádiz, or school was unfortunate.

"Ana is helping me a bit but I need to speak it more throughout the day. Unfortunately, everyone here wants to practice their English."

"You should come to my place and meet a few of my friends," she winked at me, taking another drag.

She had been here for about four weeks longer than me and her language skills had impressed Louis who hired her for the reception job. Like me, Kym looked for accommodations outside of the hostel, mainly because Louis required all the beds for the summer but also to get a break from him and the people there. We all got to meet Louis's wicked temper very quickly and unfortunately Kym and Lucia were usually on the receiving end of it. She had found a house near Jerez de la Frontera, a thirty-minute car ride from Cádiz.

"Sure, I would love that," I smiled.

The next time we had a day off together I hopped into her rusty jeep, and with Empire Of The Sun blaring through the speakers and the wind in our hair we left Cádiz behind us. We stopped at the store on our way, picked up several bottles of booze, cigarettes, and chips and rushed to the house to get the party started. When we arrived her roommate Thomas, a young German fellow, helped us with the groceries and opened a few beers, while Kym got started on mixing rum, Moroccan mint and two liters of 7UP in a big pot. She said she makes the best mojitos and from what I remember, I couldn't disagree.

Later, under the moonlight, while reggae tunes set the tone, I was finding it harder and harder to keep up with the conversations between her and her Spanish speaking friends. Kym would often translate, but I suppose it was getting a little tiring for her and so I resorted back to how I used to communicate with the Czech and Hungarian kids on our summer camping trips. Hand gestures and one-word sentences. Eventually, I was feeling left out. I had been watching the group light up one cigarette after another and with each mojito my urge to join them became stronger.

"Can I have one of those?" I finally broke down.

Kym grabbed the pack off the table and with a light tap and tilt, she freed one away from the rest and pointed it to me. There was a minor hesitation as I reached for that perfectly rolled stick of tobacco. In a flash I heard my councillor whisper in my mind, one away from a pack a day, and with a dismissive eye roll I leaned into her flame. I inhaled deeply. Oh God. *How can something so gross feel so right?* I wondered.

As I boarded the catamaran from Puerto de la Santa Maria back to Cádiz the next day, I still had Empire Of The Sun ringing in my ears,

"Walking on a dream
How can I explain . . .
We are always running for the thrill of it, thrill of it
Always pushing up the hill, searching for the thrill of it"

It was a great time and I felt uninhibited and part of a group of free spirits. I didn't like that the price for feeling like I belonged was smoking. And yet, as my other faculties were slowly dwindling with each sip of my mojito, it became the ultimate affirmation that said, "I'm like you."

How could I let this happen? A familiar nagging in my head began to rant. *Four fucking years! Don't you remember how excruciating it was to quit? Now, you're back to square one.* But then I heard my uninhibited voice say, *Oh Yvonne, stop beating yourself up over it. It's summer! You're experiencing the wild. Let loose a little. You're too fixated on being perfect. This is freedom!*

Only it wasn't. I couldn't stop agonizing over letting myself down, and I wanted someone to tell me that it was OK. From the beach one afternoon, I called my friend Siobhan in Canada. She was just starting another day at the office.

"I started to smoke again," I blurted out as soon as she answered the phone.

"OK, so what?"

"I worked so hard at being a non-smoker and I loved not having to worry about this fucking habit. Now I'm back to thinking about it every waking minute. Please tell me that it will be OK, that it's only temporary. Just a summer thing, right?" I knew I was asking the impossible of her but I didn't know what else to do. I needed relief from the ping-pong game in my head.

"I'm sure you'll be fine once you get this out of your system, Yvonne. Have fun and don't worry about it."

I hated having to phone home to get help on something I thought I was done with. Siobhan never even knew me as a smoker. I wanted to be able to share my evolution and how enlightening it was. How I was finding myself, like Elizabeth Gilbert was in the ashram in India. Instead, this simple act of 'just one' catapulted me so far back in my mind that everything I had gained from being a non-smoker, the lessons I learned from quitting this habit all the way to boarding a plane to Europe by myself and my deeply insightful experience on the beach in Tarifa, was somehow all demolished.

Freedom Seeker | 47

With each day that I lit up, my sense of self-worth faded considerably and my need for validation and love grew. I was so twisted up about this that somehow I believed if I drank I could blame my inability to stop smoking on that. Soon I needed a smoke to wake up or to get a break from sweating profusely cleaning bathrooms and changing beds. I needed one as a digestive after a meal or to help me celebrate another completed shift at the hostel. If Louis yelled at us I needed a smoke to calm my nerves, and when Lucia needed consoling for her bleeding heart I smoked quietly beside her thinking of how to comfort her. It was my currency to friendship and belonging. It was there for me when I was feeling sad, when I missed home, and when all of us gathered on The Melting Pot terrace getting primed for yet another night out dancing. Most of all, it became my beating stick. A way I could punish and reassure myself that I wasn't enough. I played that game of quit for a few weeks but eventually surrendered to the fact that I was a smoker again, at least while I was in Spain. *I'll quit the day I leave here,* I promised myself. I grabbed my green shoulder bag and stepped out into the hot August sun. It was my day off, and a few girls I met on a night out on the town invited me to join them at the Camposoto Reggae Jam in San Fernando.

We met at the Cádiz train station at two o'clock. I stopped at the nearby convenience store and picked up snacks, water, and a few bottles of beer to share amongst us. Playa de Camposoto was a short forty-minute train ride, south of Cádiz. When we arrived I gasped at the site of the endless, white, undeveloped sand beach. We found an unoccupied spot a few meters from the water and immediately sprawled out on the big beach blanket one of the girls brought. I could hear the distinctive reggae beat from the park behind the sand dunes, and I let the gentle rhythm of the sound sway me into a lull. I closed my eyes and let the sun kiss my face, wishing I could forever remember the weightlessness of not worrying about a thing.

For my final days in Spain, and to celebrate my thirty-third birthday, I decided to take a trip back to Tarifa to retrace my steps along the beach and reconnect with the energy that had enabled me to see clearly into the future and what I wanted to create. I called Rafi and booked two nights in The Melting Pot before I hopped onto my Samsung Notebook to search out a company that could make another bucket list item a reality.

Arriving in Tarifa late morning, I wasted no time. I followed the directions down the narrow street and found the address I had scribbled on a piece of paper. A young Andalucian man opened the door, looking as bewildered as his black hair was.

"*Hola! Soy* Yvonne. I'm here for that horseback ride *por la playa*," I mumbled in broken Spanish.

He shook his head and pointed toward the beach. Evidently, I was at the wrong meeting spot or the group had already left. I couldn't understand his thick accent. I hung my head in disappointment, but just as I was about to turn away, he grabbed my arm and motioned his head to follow him into the courtyard where he hopped on a white Vespa and pointed to the seat behind him. We cruised swiftly through the streets and to the nearby stables where a few smaller horses were huddling in the shade of the giant eucalyptus tree.

"*Gracias*," I said with my hands folded in front of my chest, bowing my head.

My horse was still saddled up but the group, if there was one, had already left. After we got the paperwork out of the way releasing them of any liability, the guide helped me onto the horse, a gentle one he assured me, and sent me on my way. I gave him a confused look, had he not understood that I'm not an experienced rider? Sure, in my dream I was galloping along the edge of the beach with water splashing up my wind-blown white gown, but in reality I needed a step stool to mount this animal and had no idea how to make it stop should we ever get into a gallop. He didn't seem concerned at all and gave the horse a slap on its muscular hind.

"Show the horse who's boss," he laughed.

And off we trotted into the sunset.

4
CRASHLANDING

Greece, October 2010

I arrived in Athens just two short hours after boarding Aegean Airlines in Rome.

Stepping through the big sliding glass doors into the hot, humid air, I immediately regretted wearing jeans. After feeling nothing but light cotton on my body for the past three months, jeans felt very restrictive. The damp fabric and a stuffed backpack weighed me down. *Has it always been this heavy?* I wondered adjusting the straps. Airlines allowed one piece of luggage within twenty-five pounds. Anything over that was an additional fee I didn't really have to spare. So, on travel days, I always wore the heaviest clothes I owned to avoid paying extra. I looked around and just like my Lonely Planet guide said, I spotted the number three bus that took me and my few belongings to the hostel in central Athens that I had booked before I left Spain.

I loved how fast scenery, language, and culture changed in Europe. From the beginning of this trip, my intention was to make my newly developed enthusiasm for the great outdoors a part of this experience. In Ireland I hiked the rolling hills of Dingle. I walked the streets of London and toured Rome by foot (the best way to see it). I hiked the serrated mountains, Monserrat, in Catalonia and wandered for miles along the Spanish beaches. Now that I was finally in Greece, naturally, I wanted to climb Mount Olympus and say hello to the gods. I was an intermediate level

50 | Yvonne Winkler

hiker with many day hikes in the Rocky Mountains involving no more than a two-thousand-foot elevation profile. Mount Olympus rose 9,500 feet from the Aegean Sea, with fifty-two separate, snow-capped peaks, making it the second tallest mountain in the Balkans. Let's be clear, this was not going to be a day hike. The full trek would take two to three days, with two additional days there and back.

Like most of this journey, my planning was vague but particular about what I wanted to see before I went home. In the weeks leading up to my departure from Cádiz, I began my research on how one would go about summiting this majestic rock that Zeus made his home. I learned that in October hikers can expect fast changing weather conditions including snow. This was creating a little bit of a dilemma for me as I had left my warm jackets and hiking boots at my aunt's house in Germany to lighten my load while I was traveling through the hot Mediterranean climate. A detour back to Germany would have cost me several hundred euros and pushed my timeline to visit Greece further into October than I liked. I decided to not worry about the weather conditions until closer to the dates and then check in with a mountain guide company when I arrived.

A big advantage of traveling in the shoulder season is that many students and other travelers had returned to the real world, leaving ample accommodation in the otherwise overcrowded hostels. It was midafternoon when I entered a small room with two bunkbeds, the curtains drawn to help keep out the heat. I sat on the lower bed and unbuckled the heavy pack. I needed to get out of those jeans and into a fresh shirt and shorts. The tiles cooled my bare, swollen feet as I unzipped the day pack from the rest to trim its width and fit into the locker. Reducing my living space to a single bed and locker after enjoying the luxury of a private room was an unpleasant adjustment. *I hope my roommate is cool at least,* I thought. Reading the title of the romance novel on the other bed, I assumed it was a woman. She too had chosen the bottom bunk, likely for the same reason. It could get really disorienting after months of traveling if you had to get up in the middle of the night to go to the bathroom. I hit my head on low ceilings several times or slipped on those oddly spaced rungs.

The door swung open, and a twenty-something year old woman with dark, dripping wet, curly hair entered the room.

"Oh hey, I'm Haylee," she said in a thick Australian accent, throwing her grey and orange toiletry bag on her bed.

I smiled, relieved, and sat cross-legged on my bed, ready to shoot the breeze for a bit. She had been traveling for a month on her own through Hungary, Poland, and Germany and finished her tour with Greece before heading back to start a new career. I immediately liked her quiet adventure spirit.

"Want to join me for drinks on the rooftop later?" she asked. "I heard it's pretty cool, and you can see the Acropolis from there all lit up at night."

"Sounds like a plan. Have you seen Athens already?"

"No, I'm going on the walking tour later today," she added.

Many hostels offered a sightseeing walking tour, which were usually led by locals or university students working for tips. I found it to be the least expensive and fastest way to get to know a place. While they too hit the "tourist" spots, they were more interested in telling the stories that the double-decker tour buses wouldn't. The juicier the story, the more authentic the experience, the higher the tip.

"Yeah, I wanted to check that out as well," I looked at the time.

"First, I need to make a call about climbing Mt. Olympus," I excused myself.

A friendly man with broken English answered on the second ring.

"Do you have winter hiking gear?" he asked.

"Not really. I bought a fleece before I left Spain, and I have day hikers and thick socks. All of my winter clothes are in Germany," I said. "Do you rent gear?"

"We don't and," he paused, "it snowed last night, and the winds are gusting. You'll need proper equipment to hike the mountain. I'm sorry I can't help you more."

The trek itself wasn't that difficult, but many mountain rescues were due to underprepared hikers. I knew that from my hikes in the Rockies.

"OK thanks," I pressed my lips together and rubbed my sweaty forehead. This had been one of my number one reasons for coming to Greece. I was so torn. I was on a shoestring budget and didn't want to make the detour to Germany to grab my boots, nor could I really justify buying new gear for this, albeit legendary, one hike. *But will I get another chance to do this?* I wondered. I had talked about summiting Mt. Olympus for months; I was so close to checking this off my list. *Would I regret not doing it over a few hundred* euros? With my head hanging low, I walked away from the front desk into the streets of Athens.

The first stop of our walking tour was the Greek parliament in Syntagma Square. Our guide, Dimitri, informed us about the riots that had been going on, which made me withdraw a bit further to the back of the small group. I had nothing to do with the EU or its austerity cuts, but I was blonde, blue-eyed and carried a German passport in a country with enormous amounts of animosity directed at Germany. Greece, much like myself, was going through hefty growing pains. When communism collapsed across eastern Europe in 1989, the so-called 'single market dream' was accomplished by the movement of goods, services, people, and money across the borders of Europe. Twenty years later, eleven new countries had joined the European Union, which was in a major debt crisis. Some eurozone members like Greece, Portugal, Ireland, and Spain were unable to refinance their unprecedented government debt that was assumed when the euro was introduced. The EU helped several countries, including Greece, confront their difficulties which provoked anti-austerity demonstrations, fighting plans to cut spending and raise taxes in exchange for a hundred-billion-euro bailout by the EU.

As I stood in the plaza I felt the anger of the Greek people, and I empathized with my countrymen and the frustrations they held as they lost almost fifty percent of the value of their livelihood overnight due to the equalization this 'single market dream'. We had seen similar hostility on a smaller scale when Germany was united. There were many from the wealthier west who bucked against federal support spending to upgrade and infuse vitality into the Eastern Bloc states. Like the people in Greece we simply wanted to better our situation, but we couldn't do it entirely without some equalization of wealth. But to lose half of their hard-earned retirement savings wasn't an easy pill to swallow either. *I guess it's always the people who have the least say over these decisions that impact their lives so fundamentally,* I thought looking at the Old Royal Palace.

"Next stop, the Temple of Zeus," Dimitri said with a raised arm, pointing up.

Oh good! I thought, if I can't reach the gods via Mt. Olympus, at least I get to see a homage to Zeus. I had held a fascination with Greece, Greek mythology, and the crystal clear, teal blue waters of the Aegean Sea ever since a young age when my aunt gifted me the book *The Odyssey* by Homer. Now that I had seen the Colosseum of Rome, I wanted to explore the threshold between the mortal world and the gods.

As we approached the site and I could see the colossal columns, my heart began to beat so loudly I could barely hear Dimitri explaining the site. Closer, I spun on my heel, and turned around again, anticipating the rest of it to mysteriously appear from behind the city's dry and dusty air. All that remained standing of the once magnificent and largest temple in Greece were twelve of the original one hundred and four columns. Twelve!

"This sucks!" Haylee whined. "I expected it to be a little more, well, magnificent."

"I'm bummed out too," I empathized, "at least we still have the Acropolis to look forward to."

"Yes!" she said cracking a smile. "I hear it's impressive."

"Not to be missed is what I read," I said in as cheerful a tone I could muster given all the disappointments so far.

Without spending any more time at the ruins of the Temple of Zeus, we made our way through a bustling market and up the hill toward the Acropolis. Sweat rolling down my neck and a little out of breath, I turned to look back at the white buildings, some garnished with pink flowers, and I wondered for a moment how I intended to climb Mt. Olympus. Nearly three months of lazing on the beaches of Spain, drinking claras and smoking Marlboros wasn't exactly the conditioning one requires to go on a three-day trek. But I quickly forgot about my agony when I saw the barricades around the Acropolis. *What's going on here now? I wondered.*

There is something so very disappointing about traveling thousands of miles to see something, knowing it will likely be the only chance I'll have to see it, to be met by a padlocked entrance and angry protesters refusing visitors to enter until their demands were met by their government. It seemed everything I had set out to do and see in Greece fell apart. *Was this place rejecting me?* Maybe it was time to go home. Standing on the dusty ground at the gates of the Acropolis, perspiring and dejected, my eyes scanned the crowd. I needed a smoke!

As Dimitri rambled on about the Parthenon, the Temple of Athena Nike, the Odeon of Herodes Atticus, and whatever else I was not about to see, I approached a small group of young visitors also coming to terms with the situation before them.

"Excuse me. You speak English?" I asked in a hushed tone like I was about to do something strictly forbidden. Determined to leave my smoking relapse in Spain, I had stopped buying cigarettes when I left Cádiz two weeks ago. "Can I please buy a cigarette off you?" I mimicked a smoking hand to my mouth and held a fifty-cent coin in the other just in case they didn't understand. The young woman squinted as the smoke from the cigarette in her mouth caught her eye and, like a pro, she handed me one and took the money.

"Gracias," I said, still in the habit of speaking Spanish. I took a long drag and blew the smoke out slowly, like I was deflating my plans for Greece. Turning to Haylee I said, "Wanna go to Santorini?"

By not doing the five-day, six-peak round trip on Mt. Olympus, I had enough money and time to go to Santorini, a place some believed to be the original home of the lost city of Atlantis, and my hope to restore my fascination with this country's stories and mythology, rather than face the modern-day politics

***

From the ferry I looked up the steep rock face of Santorini, rising out of the turquoise waters, and wondered if we'd have to take those gnarly stairs that clung to the whitewashed buildings. Swarms of swallows flew playfully around the rock where insects were most abundant, and seagulls circled the blue domes I recognized from the Lonely Planet guide pictures.

Haylee and I had decided to stay on the more affordable east coast of the island where we each had our own room, a little kitchenette and balcony facing the courtyard pool. After nearly eight months of sharing kitchens and showers, this was luxury. I was craving dominion and self-reliance. I filled the mini fridge with some yogurt, fruit, and lettuce, ambitious to get back into the swing of a healthier lifestyle. Bottles of sparkling water replaced beer, and packages of spearmint gum would help with the cigarette cravings. At night I sat on my balcony with my Notebook computer and a freshly made bowl of salad, occasionally looking up from writing about my travels, and gazed at the last stragglers getting their party on at the tiki bar.

I glanced at the time on my computer and counted eight hours backward. It was too early to Skype with anyone in Canada. They were all in the middle of their workday. There it was again. The guilt. After all this time, I still felt guilty for gallivanting through Europe while my parents, especially, and friends were drudging away at their jobs. I opened Facebook and scrolled through my feed, trying to connect. When that failed, I attempted to type a post that described how, after eight months of seeing new places, I still felt lost. I was tired and lonely. This struck me as weird since I had spent so very little time alone. Surrounded by people all the time, doing what many would consider a dream. What was missing? I was longing for a partner, for a place I could unpack my bags and stay a while. I was craving pumpkin pie and sweater weather. Like the land around me, I was parched for rain. I missed home.

I checked back on my post to see if I hooked someone into a dialogue. A few gave me the thumbs-up and one comment. A former colleague who was on maternity leave responded with a sour note about being a crybaby. I knew she was hardly the ideal person to relate, but I took it down immediately because seeing more of the same would only make me feel worse. They were right. I knew I was blessed to be traveling and seeing the world. *Why was I still not happy?*

Unable to motivate myself to get out of bed, I spent the next three days locked in my room. Simultaneous eruptions of grief, anger, and shame had me pull the covers over my head and binge watch *How I Met Your Mother*. Even a room all to myself wasn't solace enough, I needed to hide away even further in a cocoon of blankets. Months of navigating foreign languages, unable to do as simple a task as ordering a burger, forging my way through cities and countries on trains, planes, and busses, always watching my back for danger and easing into "fun" had left a huge impact on my nervous system. I was burned-out on the inside as much as my skin was on the outside from the endless days of summer. I needed the natural transition into respite that fall brings, only the temperatures weren't dropping and there were no trees with leaves changing from green to orange, yellow, and red in Greece. The sole indicator here that winter was coming was a lack of tourists flocking to the beaches due to the increasing intensity of the waves pounding into the black sand.

But going back meant that my sabbatical was over. *Was I ready for the real world again?* As much as I longed for the familiar routines of fall, I wanted to stay in Europe for as long as I could. I felt I hadn't finished what I came for. Every day I checked with the front desk about the *Meltemi* that was coming our way. *Meltemi* is the northerly wind that creates harsh sailing conditions and can last for several days. Not having the most robust stomach for boats, especially on large swells, I had to be ready to leave any day to avoid being stuck on the island until spring. I considered it for a moment, *I could open a bed-and-breakfast and live a simple life by the sea, like Meryl Streep's character in Mama Mia.* Only I wasn't pregnant, nor did I have any prospects, so I would still be alone.

It was clear that I was done with traveling through places where I didn't speak the language, sharing rooms with strangers, and living with only one change of clothes. My hair, dry and brittle from the salt and sun, sorely needed some love, and my body required nourishment beyond white bread and strawberry jam. The desire to create something meaningful with my life was getting louder by the day, and yet I was terrified of losing the simplicity my life had now and trading it in for the rat race, ambition, and overwork. I knew that this life wasn't going to be enough for the seeker in me but going back to my old life wasn't an option anymore either. I loved the freedom that came from the detachment of things, titles, and expectations, but I couldn't deny that being able to afford my own place and regular hair care were important to me. Besides, I wanted to make an impact in the world and that required resources. Aargh, the ping-pong game in my head made me dizzy.

In an attempt to distract these never-ending thoughts, I thumbed through the travel guide on my bedside table and learned that Santorini, once a big, round island, was now one of five separate islands due to the former intense activity of the now mainly underwater volcano. They offered tours to see the crater on one of the islands, only a fifteen-minute boat ride away. *Hmm,* I thought. *I've never been on or in a volcano.* Bright and early the next morning, I took the bus to Fira Theotokopoulos Main Square from where I carefully scaled down some five hundred and fifty slippery, cobbled stone stairs littered with donkey poop to the old port at the western edge of the island, where a gorgeous traditional wooden boat waited to take a handful of us across to volcano bay.

The island was the crater. I walked around on the black, lifeless ground cautious to not get too close to the volcano's steam vents. *What if this suddenly erupted? Could I say I've lived my life to the fullest?* I pondered. *Nope, definitely not yet.* A sudden charge of urgency to get living and stop waiting for life to happen caused me to stand up a little straighter. I turned to the woman standing next to me and asked, "Could you please take a picture of me?"

When the captain dropped us back off on Santorini, I glanced up the winding cliff path I had descended earlier. I looked over to the donkeys that huddled together under the small piece of shade cast by the rock wall and decided to take my chance with one of them. *The locals used them as a mode of transportation for centuries, it couldn't be that bad, could it?* I reasoned, always looking for ways to simplify the steps and avoid slipping on donkey poop.

When we finally reached the top, I had a new understanding of the phrase 'stubborn like a mule'. Turns out, you can't rush a donkey to do anything it doesn't want to do. The jackass I rode must have sensed my fear of falling down the steep cliff to my inevitable death and decided to take it real slow up the slope. The instructions were in Greek, and I was too nervous to jam my heels into its ribs, certain it would buck me off in protest.

A growl from the pit of my stomach reminded me that it was lunchtime. To my right was a restaurant with a beautiful balcony overlooking the Aegean Sea and the volcano. A small table for two, with a white linen tablecloth and a single flower in a ceramic vase invited me to have a seat and enjoy this view for a little bit longer. I decided to treat myself to a more formal dining experience than the usual, inexpensive street food. I scanned the menu of mouth-watering images and spotted my favorite dish, souvlaki with freshly made tzatziki. The waiter brought me some sparkling water with lemon, and I sat back and took in the panorama of white buildings and cerulean waves. I heard a couple behind me, just coming up the donkey trail, breathlessly discussing what they should do next before they had to get back and board their cruise ship, which from where I sat looked like a small toy. *I wish I had a partner to enjoy this experience with. What would we do next?* I wondered dreamingly.

I had been without a man for nearly eighteen months now and while it was great to do my own thing without a care about anyone else, I was missing a partner to share moments like these with. Secretly, I had hoped that a handsome summer romance would turn into more. Unfortunately, there was no summer romance as I was never particularly good at playing the dating game and flirting with men had always been awkward and weird. Like many things in my life, I didn't really know what I wanted until I had it and realized that wasn't it. I'd take what was handed to me or dated whoever pursued me. On the few occasions I fell head over heels for someone it looked more like stalking. I would obsess about his every move and whereabouts so that I could accidentally bump into him. I wasn't raised with a healthy expression of desire, nor with the gumption to go after my dreams. I learned that discipline and perseverance was how we got things in life, and I somehow applied that philosophy to my romantic interests as well.

60 | Yvonne Winkler

Maybe my first love was to blame. He was the only man I wanted and in a terrible twist of fate, our young hearts were broken not once but twice. The first time, we had left my hometown in such a hurry that I couldn't even say goodbye. He heard the news through a friend. It was devastating but we reconnected about six months after our move to West Germany at a teenage disco club in my old hometown and stayed in touch via handwritten letters. We saw each other whenever I visited my grandmothers but after about a year he told me that he couldn't do this long-distance relationship thing and he didn't want to see me anymore. My heart broke into a million little pieces on my grandma's bathroom floor. I can still see her standing in the doorway, helplessly inquiring what had happened that had me so undone. He was that first love that ended up being the last.

Had I somehow denied myself to experience love again? My heart often caught between what I really wanted but not able to bare the loss or rejection. I had certainly guarded myself from rejection for years and didn't allow myself to get attached to anyone. I took another bite of my souvlaki. *Have I been doing this with other desires in life as well?* I pondered. I put my knife and fork on the empty plate and looked over the wide-open space in front of me. *If my dreams didn't become real then I wouldn't have to feel the pain of the loss either.* I was playing this out in that moment! I was in a place I had always wanted to be, free as a bird to go and do whatever I wanted, yet I denied myself to be in the moment of it and instead made myself miserable and depressed.

"No more!" I said out loud to everyone's surprise. I got up, excused myself by mumbling something about the sun getting to me, put €10 on the table to cover my meal, water, and tip and rushed back to the hotel.

Haylee was waiting for me at the pool bar and waved excitedly when she saw me.

"Hey, where have you been? I've been looking all over for you," she said a little concerned. "I'm thinking of renting an ATV

Freedom Seeker | 61

and cruising the entire island before I leave tomorrow. You wanna join me?"

"Absolutely! I'll just change and meet you there in fifteen."

I put on my skort and a fresh tank top and walked to the ATV rental place that was right around the corner from our hostel. After a short intro on the breaks, clutch, and horn we were off, each on our own four wheels. Truly, there was nothing like the humming of a 570 volt engine between your legs to feel powerfully independent. *Who needs a guy?* I thought as we cruised the very narrow streets of Oia, Akrotiri, and finished with Fira where we watched the spectacular sunset. It was the perfect way to round off my stay in Greece, and the next morning I boarded the ferry back to the mainland and then a plane back to Germany. I still wasn't sure if I was ready to return to my life in Canada yet. Hearing my mother language, seeing myself reflected in quirky, familiar traits in my relatives, and sitting on my grandmother's couch gave me a comfort and a sense of belonging I didn't realize I was missing. What if this was all I needed to be happy?

***

Germany, 1992

It didn't take me long to immerse myself in the community of our new home, Henfenfeld, a small town of almost two thousand people nestled in the Alps. In fact, my mom and I loved it so much that we rarely thought about the home and friends we left behind in East Germany anymore. It was a bit awkward for us to adjust from our house and big backyard to a second floor, two-bedroom apartment with no garden, not even a balcony, but we were happy to have a place of our own again. Dad found work as a door-to-door insurance salesman and spent most of the week on the road, and Mom worked as a tailor in an upscale women's boutique in nearby Nürnberg. I had just started the German equivalent of high school in Hersbruck, a melting pot for all teens from the neighbouring communities which expanded my social network considerably. Unlike former communist Germany where we attended one school from grade one through ten, here I had to decide in grade six what career path I thought I might want to take and choose between three educational systems, *Hauptschule*, *Realschule*, or *Gymnasium*. All kids attended *Hauptschule* or main school until grade six. Students who wanted to extend their education, learn at least one other language, and sought middle class careers then switched to *Realschule*, the equivalent of high school. The especially bright students were prepped for a post-secondary degree and they attended *Gymnasium*. I chose the middle of the road. The East German curriculum was advanced and made it easy for me to integrate into the Bavarian school system which was ranked the highest in Germany. The only subject I was behind on was English. I had learned Russian as a second language and needed to work with a tutor after school to hone the English skills that I missed.

In a mere few months, I was on the student council, a basketball team, fell in and out of love at least three times, and some Saturdays I went to football games with a friend and her posse. Life was good and busy. I fitted in with everyone and no one and that was perfectly OK with me. My interests were as diverse as the people I surrounded myself with and my chameleon-like personality allowed me to slip into each social setting with ease. I hung out at tennis clubs with the smart uni clique and my boyfriend Sebastian, and sometimes I drank tea from handmade clay mugs with the people I met in main school when we first moved into this community.

The people I spent the most significant amount of time with was 'the gang'. A diverse mix of girls and boys from all educational backgrounds and ages, brought together by our common interest in the opposite sex, cigarettes, and hair bands. We hung out on the steps to the school during lunch and met after school at the town center water fountain until dusk before catching the last bus home.

At fourteen, I thought I knew everything and living in a free world, little stood between me and my independence. Nothing made that point as obvious as a pack of Marlboros in my jean jacket pocket and a black pair of Doc Martens. I let my hair grow out from the choppy childhood pixie cut to a mid-length bob. I wore oversized, plain white t-shirts, stonewashed jeans and I had a wide collection of bandanas, the most badass accessory of the 90s, thank you Axel Rose.

I met Carola on the first day of *Realschule*. She and Sabine were from the same town and sat at the desk behind me and Doris, who I knew from grade six. Carola, unlike Sabine, was a quiet observer. At lunch break, she tossed her crinkly long hair and decidedly wove her arm through mine as we strode to the school cafeteria to share a piece of freshly made caramelized onion bread. From that moment forward we were inseparable.

She lived in another town, too far for a quick bike ride, so we spent every moment together in school and another hour on the telephone once we got home. We talked about the tumultuous times we faced in school, with boys, and our families. There wasn't anything I couldn't tell Carola and she was always the first person I shared all the details of my life with.

My dad, eager to make up for lost time, quickly climbed the ranks and became an insurance branch manager. As a reward he upgraded the used, silver blue Volkswagen Golf to a brand new, midsized Opel Vectra and he bought Mom a chili red Opel Corsa hatchback, perfect for zipping around town and getting groceries. As a family, we spent many weekends exploring our new surroundings, cycling through the picturesque Franconia, going to the world-famous Christmas market and tasting our way through Greek, Italian, and Chinese restaurants. Mom and I loved to craft and were enraptured by the selection of art supplies around us. On rainy days, we made hand-painted masks, brooches and re-established my grandmother's favorite handcraft, knitting.

64 | Yvonne Winkler

Our hunger for all things, from food to fashion, shopping, travel, music, and entertainment was almost insatiable and we didn't waste any time. We drank the most expensive coffee in Venice, kissed Romeo's hand in fair Verona, took the Glacier Express from St. Moritz to Zermatt, and took an enchanting drive over the San Bernardino pass in the Swiss Alps. In Austria I skied, strolled through romantic renaissance courtyards and marveled at cathedrals. The biggest surprise was when Dad bought us tickets to the then undeveloped Dominican Republic one Christmas holiday. I had never been on a plane or seen palm trees. I had also never been exposed to life outside of Europe. It was a nightmare of lost luggage, sea sickness, and the inability to communicate properly, and yet it was a taste of privilege and I was hooked.

Through the insurance team my dad was part of he acquainted a guy, René, who, like us, came from former communist Germany and had a fierce hunger for freedom. They spent much time making plans about a better future which looked a lot different for René, a single man and a few years younger than Dad, who always had us to think about. As the two of them took the insurance industry by storm, they began to set their sights on bigger dreams.

On the night of my dad's fortieth birthday, the three of us dressed up to celebrate this milestone in his favorite restaurant. Over Peking duck and crispy fried rice, he casually mentioned that he and René had been talking about a trip to Eastern Canada to explore. They heard that they could still buy land there and wanted to see for themselves if there was a viable investment opportunity. Busy with my own life and the latest boy drama, I paid little attention to where and why until he returned from his trip into the woods.

After a long day at school, followed by working at the local grocery store stocking shelves, I had just switched off my bedside lamp and closed my eyes when I heard a quiet knock on my door. It was Dad who poked his head in to say hello and goodnight.

"We rushed back from the airport in the hope I would catch you before you're asleep," he said winded from running up the stairs.

"*Vati!* How was your trip?" I propped myself onto my elbow and turned the light back on.

"It was breathtaking," he snuck in and sat down on the edge of my bed. "I'll tell you all about it tomorrow, OK?"

"And?" I queried impatiently. "Did you find a house for us so we can visit in the summer?"

"Yes!" He snickered. "Sort of. Actually, we bought a hotel," he paused and scanned my face for a reaction. "It's a Canadian log house in Nova Scotia and I'm going to move there to run it."

"Wait what?" I sat up and propped my pillow to support me for this obviously longer discussion.

"Did you just say that you're moving to Canada?" I shook my head in disbelieve. It couldn't be. "What about Mom? What about me?"

"Well, I'm still warming your mother up to this idea. I would go alone for now and get things set up and running. You and Mom come in the summer to visit and decide then if you want to stay." He handed me a miniature grey seal souvenir and continued, "It's real seal fur. They are everywhere on Cape Breton Island."

I petted the souvenir and was surprised by its comforting softness. *It would be cool to see seals and other sea animals in real life,* I thought. My brain was working hard to interpret everything my dad had just dropped on me, and in that moment it only registered one question. The one I knew my dad could not say "no" to if he wanted me to buy into his plan.

"Can we then finally get a dog?" I had been asking for a furry companion for as long as I could remember but my request was always denied.

66 | Yvonne Winkler

"Yes," he grinned from ear to ear, "we'll get you a puppy." He kissed my forehead and squeezed my hand as to seal our secret pact and left my room.

I laid there wide-awake for a long time, thinking about what kind of dog I would get and wondered what living by the ocean would be like.

Over the next several months, Mom and Dad spoke very little with each other and when they did, it quickly turned into arguments. I hid away in my room, listening to my favorite new CD, *Waking Up the Neighbors*, or hung out at Carola's place distracting my mind with crossword puzzles. Mom wasn't on board with any of the options presented to her. She was angry that Dad had made such an important decision that impacted all of us, without consulting her first. In a desperate attempt to either fall in love with Dad's dream or to talk him out of it, she agreed to fly to Nova Scotia with him to see for herself what he had committed their life to.

It was so surreal and, being a typical teenager, I only thought about how this inconvenienced my life and how I had to say goodbye to my newly established life and friends yet again. Grandma stayed with me while Mom and Dad were in Canada but I was old enough to hang out with minimal adult supervision, and I spent most days escaping the realities of my impending future by hanging out with my friends, smoking cigarettes, and drinking radlers, a refreshing beer and soda mix.

When Mom returned from Canada alone, I knew we were moving again.

"Dad decided to stay and get the lodge ready for our first summer season," she said strained. "He said that he wants you to finish your school year here and then you'll come over with René."

"And what about you?" I asked holding myself up against the kitchen wall where Mom was making a fresh salad for dinner.

"I have to get back there and help set up the restaurant and hotel." Her short replies spoke volumes. It was an impossible decision for her to make. If she stayed, she lost her marriage, her life partner, and possibly me. If she went, she faced saying goodbye to everything she knew once again. She had to learn a new language, culture, and be far away from her sisters, brother, and aging mother. "I don't know the first thing about the hotel business," she cried. "What was he thinking?"

I didn't know how to comfort her as it felt like a betrayal to my dad. I didn't want to have to choose between them either but mostly I wanted to believe that Dad knew what he was doing and it would all work out.

As the days got longer and the sun dried up the last of a rainy spring, all the towns around us were getting ready for summer festivals and beer gardens. Carola and I sat on the stairs after school and planned the weekend. Our biggest obstacle was how we'd get home from the parties because the last train left town at ten thirty. Puffing away, the butts piling up at our feet, we finally came up with a sure-fire plan.

"I'll tell my mom that I'm staying at your house, and you tell yours that you're staying with Sabine." I said proudly. "Sabine, you tell your dad you're staying at Carola's."

"In the morning, we each take the train home and our parents will never know we stayed at the party." Carola echoed.

We parted ways and agreed to meet up at seven.

I came home to shower and change clothes and found Mom bent over moving boxes packing up the contents of our living room.

"Can you give me a ride to Carola's tonight?"

"Yvonne, can't you see I'm up to my elbows in packing?" she said exasperated. "I would really like you to stay here and give me a hand. I can't believe both you and your father are leaving me to do all of this by myself." She choked back her tears and quickly turned back to carefully wrapping the handmade Italian crystal wine goblets she bought in Venice a year ago. Guilt crept up the back of neck.

"Mom, it's Friday night and I won't have much more time with Carola," I begged. I would have walked, and while the community we lived in was safe, it was common sense even for me to not walk alone at night to the next town.

"Fine!" Mom hissed. "But tomorrow you help me with packing!"

Two weeks later, my mom left Germany with two giant suitcases that were as heavy as her heart. My grandma came to stay with me again until I finished eighth grade at the end of July. Carola slept over almost every day so we could squeeze in every last minute of time together. We slept on a bed of blankets on the floor of my otherwise empty bedroom. Mom had sold almost all our furniture and stored some of our personal items at my friend Sabine's family property until we could arrange to ship them across the Atlantic. The windows were wide-open to clear the blue smoke and stench from an overflowing ashtray after another late night of talking about this guy Carola was in love with and how stupid he was for not noticing her. Neither of us touched on the topic of life after I'd be gone and yet it was painfully lingering heavily over our heads like the clouds of smoke.

5
WALKING ON FIRE

Nova Scotia 1993

René, who had by now become a family friend as well as my dad's business partner, chaperoned me on this trip across continents as I was only fifteen, and direct flights were not possible from Frankfurt to Halifax at the time. When our plane touched down at Toronto Pearson Airport, sometime in the late afternoon, all my tired body craved was a cigarette. Unfortunately, my two big suitcases and grey backpack, that was about the size of me, sparked the Canadian customs and immigration officer's interest and delayed any relief as they led me into an interrogation room. Going through security evoked a lot of memories. My chest was so tight, I had difficulty breathing, and abrupt hot flashes tricked me into feeling I was suffocating. As the agent across the small table from me fired question after question about where I was going, how long I was staying, and on and on, the fateful day on the side of the road in the middle of Hungary flashed before my eyes and made me stumble on the few English words I knew. This raised even more suspicion about a fifteen-year-old traveling across continents without her parents; they strip searched me, for drugs I assumed. Whenever we crossed a border in former East Germany, my mom told me to lay down in the back seat of the car and go to sleep. In the same way, I tuned out the intense stress that arose from this interrogation by detaching from my body. Checked out from any

69

emotions, cold and numb to the fear inside. I couldn't show any weakness to the officers, just like my mom in that ditch.

When we finally arrived in Halifax, I was so exhausted I declined to join René for a bite to eat and headed straight to my room. We'd been traveling for over eighteen hours and decided to check into a hotel for the night before we'd continue the four-hour drive to our final destination, Cape Breton Island. Grateful to see the neatly made queen-size bed, my eyes immediately spotted the Thank you for Not Smoking sign and a locked mini bar. Not old enough to smoke or drink in Nova Scotia, I curled up in a ball and cried myself to sleep, half a world away from everything I knew.

For the first time since Dad imparted the news that he was moving to Canada, I wondered why. We had a good life in West Germany. Mom seemed happy and had established a bit of name for herself at the boutique where she worked. Dad was earning more money than he ever imaged and afforded us privileges we had never had before. I had more friends than I thought, given the parade of people that came to the house over the last few days to say goodbye. They brought me framed pictures of us together, teddy bears, and other soft animals to keep me company in a new and foreign land. Saying goodbye again to everything I knew was as heart-wrenching as the day we drove through my hometown, Tauscha, waving goodbye.

Carola was really my biggest heartbreak. She was my best friend. On our last day together, she gave me a silver necklace with the inscription *Ich vermiss Dich, Caro* (I miss you, Caro), and I quickly clasped it around my neck before tears could mess up my freshly applied make up. I could still feel her arms around me as we stood in front of my house before she turned to her mom who was waiting in the car, also crying. I heard a knock on my door, pulling me from my memories of Carola and quickly gathered my luggage.

"Are you ready to see your parents?" René greeted me with an enthusiasm I could not match. Of course, I was excited to see

Freedom Seeker | 71

my parents but the weight of my reality was making it difficult to smile. He grabbed the two suitcases, and we made our way to the rental car. The warm, salty summer air swept across my face as René stacked our luggage into the Jeep Cherokee. Everything was so different.

"Nova Scotia is Latin and means New Scotland." René practised his tourism pitch on me as I gazed out the window searching for the ocean.

We turned inland onto the Trans-Canada Highway 104, the fastest and possibly only route to rugged Cape Breton Island, which was bridged to mainland Nova Scotia via a two-kilometer causeway, massive highlands on the north side, and a saltwater lake at its core. The lodge was built by three Americans as a private hunting retreat, nestled deep into the woods with only few houses on either side. René opened the main entrance for me and as I walked down the hall, I counted eight rooms before I stepped into a cozy rustic lounge with panoramic view over the deep blue waters of the Bras d'Or Lake.

"Wow," I gasped. "This is breathtaking!"

Mom poked her head around the corner and quickly removed her apron before embracing me in a tearful hug.

"Dad will be here any moment, he just went to St. Peter's to pick up more building supplies," she took a step back and looked me over as though to make sure I didn't leave any part of me in Germany. We went downstairs into the kitchen and soon to be restaurant to meet Sharon and Elsie, two women who had managed the lodge for the Americans and now helped my parents with the transition from private retreat to hotel and restaurant. Sharon was a bit older than my mom and lived in the house next to our property. Elsie had been the lodge keeper and lived in one of the rooms during the week. On weekends she'd drive to her home in Sydney, approximately an hour and a half from the lodge.

"Dad has arranged for all of us to go out on the boat today," Mom said with forced excitement.

"Who's all of us?" I asked licking the peanut butter from the roof of my mouth. I had never had this American staple.

"Sharon and Elsie," Mom pulled out a chair and sat across from me at the table. "Sharon invited her daughter Tuesdaie who is your age. I guess it would be good for you to know someone who can show you around school," she continued. "Eddy is the guy with the boat and every boat needs a captain, René." She winked at me. We had nicknamed René "The Captain" because he was always leading the charge. He had redeeming qualities, and I could see why Dad liked him as a friend and business partner but Mom was not a fan. She blamed him for getting us into this mess and for being reckless with the lives of others. My guess was that this boat outing was also René's idea to show us all a good time and forget about our losses. He was a clever salesman.

We piled awkwardly onto the boat, and René cranked the radio and handed us a beer. Before long, on the calm waters of the Bras d'Or Lake, I had forgotten that I was thousands of miles away from home and my friends. Tuesdaie and I bonded quickly over beer and cigarettes. As the song *Should I Stay Or Should I Go* blared over the speakers, we approached Marble Mountain beach where two handsome boys caught the rope and pulled Eddy's boat to shore. Tuesdaie knew them and somehow convinced our parents to let us stay with them for the rest of the day. It was a teenage summer dream and just like Johnny, Baby, and the rest of the Camp Kellerman's crew from *Dirty Dancing*, we had the time of our life, and I could see myself fitting in just fine.

Mom and Dad were not entirely new to the hospitality business. They ran a very successful garden restaurant for nearly three years in East Germany before that pivotal summer of '89. But the rules were different in Nova Scotia and, dare I say, different for foreigners. There was a noticeable contrast in culture, work ethic, and of course an entirely new language. I had some knowledge of English from school and quickly caught onto the dialect and nuances. Mom and Dad probably could have spoken better Russian than English which left them frustrated that, for all the legal matters on getting a business started, they had to rely on the only German-English speaking lawyer on the island. Wherever I could, I used my yellow *Langenscheidt* translator pocketbook to translate the sales agreements and various business licenses word-for-word.

When September and the first day of school rolled around, I was nervously puffing away on a cigarette. Mom offered me some toast with peanut butter, but I declined. I could see the disapproving look on her face. She didn't like me choosing cigarettes over breakfast, but she had so many other problems to deal with that I guess she had to choose her battles. When the yellow school bus pulled up at the end of our driveway, a kind smile greeted me.

"Hello, you must be Yvonne. Hop on and take a seat in the back with the others," the driver said.

Tuesdaie waved at me to sit beside her, and I was relieved to know someone who could help me navigate my first days of high school. I was thrust into the grade nine classroom with the thought that I'll catch on to the language quickly, and with the help of my English teacher, a dedicated and caring educator, I picked courses that stretched my academic career and set me up for the possibility of attending university, if I wanted to, after grade twelve.

I bonded most quickly with the normals, stoners, and loners who lit up every break in a wind sheltered corner around the building and sometimes even in the bathroom. The populars disliked me because the jocks did, and the brains never really noticed me at all. It was a damaging lesson of learning where I fitted in, and three o'clock couldn't come fast enough, when I could go home to my little haven of lodge people and my golden retriever.

The local business community made it obvious that we weren't welcome. They overcharged us for materials and supplies, vandalized our business signs with swastikas, threatened my dad's life and set fire to our property. I felt the brunt of it the most when I had to see the guidance councillor, Mrs. Stone, for help with my postsecondary education choice, and she told me plainly that I needn't bother sending applications to universities.

"College is where your type goes."

Enraged, I stormed out of her dingy ten by ten office and stumbled into three of the prep girls who giggled their way into Mrs. Stone's office, slamming the door in my face.

This fuelled a determination in me. For the last term I buckled down to improve my grades. The way I saw it, university was my ticket out of this town, and a way I could prove to my dad that our move and the struggles weren't for nothing.

***

Canada, December 2010

"Welcome back!" Jennifer was smiling from ear to ear. Christmas was her favorite time of year, and she couldn't wait to gather her friends and family around the traditionally decorated tree in the house she bought with her boyfriend last summer. I felt connected to her and most of the people she invited, and their curiosity about my adventure made me feel a bit like a celebrity. It was a welcome comfort to be around familiar faces where I could let my guard down. Still, much had changed around here over the last nine months. At thirty-three, I was one of the oldest amongst my girlfriends by five years. Over sugar cookies and vodka martinis, I listened while they shared the big changes they had made.

"I gave up my bachelor pad and moved onto the boyfriend's family range," Lindsay chimed in. "We're saving up to buy our first house."

Some had changed jobs, and others were expecting their first child. It seemed everyone was moving onto the next phase of their adult life, except me. I was going backward. *Still single, living back in my parents' basement with nothing more than a pocketful of sand.* I thought as I slipped into my own bed, grateful to have at least that luxury back and a closet with a wider selection of clothes.

The next morning, I wasted no time getting to the task of finding a job because a fifteen dollar a day budget wouldn't cut it in Canada. Besides, a quick Google search revealed that the tuition for that Ayuveda training I wanted to do was going to be around ten grand. I also needed a car which required insurance and, most importantly, I had to get out of my parents' basement. I was open to finding something in Calgary, but really wanted to get back to the life I had in Vancouver before I left.

Resourceful and connected, I was on a plane to Vancouver International Airport for two interviews within a couple of weeks. One of the interviews was with the former chief marketing officer, to whom I had so boldly demonstrated my leadership skills six years earlier on the Red Mile. He had since started his own company and was looking for a hard-working, insurance trained go-getter like myself. My role would have been very similar to my old account management job.

"Of course, with more freedom and money," he added.

I liked him. He was a stand-up guy with keen business sense and excellent relationship skills. But I was concerned about getting back into something I already knew wasn't fulfilling me, although it was only temporary. I didn't want the job to drain me.

The second interview was with a certified financial planner who needed an executive assistant. It was prestigious and a little more money, but the position required a Mutual Fund Dealers Association certificate. I considered the offer for a split-second and then remembered the agonizing hours I had spent in the library studying for the Certified Financial Planner designation. My goal was to find a job that I could do with my eyes closed and that wouldn't require any new skills or energy. In my spare time, I wanted to learn all that I could about the spa and wellness industry, how to find funding, and writing a strong business plan for the wellness temple and Ayurveda center. By all accounts, that was a steep enough learning curve.

As I walked through the streets of my beloved Vancouver to Adriana's place, where she had made up her couch for me to spend the night, I noticed things I hadn't before I left for Europe. Garbage littered the sidewalks from overflowing trash cans. The smell of urine crept up my nose as I passed the side entrance of a convenience store where a shopping cart with a tarp sheltered a man from the January drizzle.

"Has the homeless situation gotten worse since I left?" I asked Adriana when I arrived at her nine hundred square foot bachelor suite overlooking English Bay.

"I think the city relocated them in preparation for the 2010 Winter Olympics, so we didn't notice them as much when you lived here," she replied in her Columbian accent, unconcerned.

It felt weird to be back here. Everything was familiar yet completely different. Over a glass of wine, Adriana told me about how she lost touch with the group we hung out with and about her weekend trips across the border to see this guy, Pete, she had met on Match, who lived in San Jose. In exchange, I tried to sum up nine months of backpacking into a few hours, something I was noticing I wasn't quite ready for. I didn't know how to talk about my experiences in a way that made sense to my friends and family, and so I often just hit the highlights and downplayed even those when I got the feeling they were asking out of politeness and not because they really wanted to know. I was lost between worlds, scared to lose the freedom and empowerment that came from navigating the world on my own terms, as I reoriented myself to where and how I fitted in.

We spent the next day like we used to, a long walk around the seawall followed by brunch at The Elbow Room Cafe where two older gay men cussed at us for fun. The Elbow Room was walking distance to the train station, where the Canada Line took me on a swift thirty-minute ride directly into the airport terminal.

Once I got settled into my seat, I put my headphones in as I had done a million times over the last year whenever I needed to shut out the world around me. Raindrops where racing down the small oval shaped window. Vancouver didn't feel as comfortable as I remembered. I pulled a pen out of my shoulder bag and folded the white paper napkin in front of me into half. On one side I wrote pros and on the other cons. I recalled how grown up I felt living here. The Saturdays I took the water taxi across to Granville Island, and the gloriously lazy Sundays reading my book on my little balcony. Something Adriana had said the night before had lingered in my mind, "I kind of lost touch with everyone." What brought us together was that we were all single. That summer, Tom, Wes, and Adriana had found love. Siobhan followed her heart to London, England, and I couldn't remember what happened to the others. I knew moving back to Vancouver wouldn't be the same anymore, and I wasn't sure that I was ready for a brand-new start. I was looking for the comforts of home and a sense of where I belonged. *You also don't have any savings left*, a quiet voice reminded me. The rational part of me knew I could replenish my finances much quicker by living in a city that was nearly half the cost of Vancouver. If I was at all serious about making my dreams of creating the Wellness Temple and Ayurveda center a reality, I had the homefield advantage in Calgary.

I spent the weekend helping Mom with the housework and dodging any and all questions about Vancouver vs Calgary from my parents as they preferred I didn't move so far away again. The choices needed to be mine and what my heart wanted. By Monday morning, with Mom's strong coffee in hand, I hit send on the two emails I had crafted late Sunday night.

"Thank you again for considering me to be part of your team Working with you would be an honor, and I know I could learn so much from you. After careful consideration, I've decided to stay in Calgary for now."

I arranged a lunch with my former manager, Cayce, for Thursday and got busy making a one-page flyer, which I framed in a simple white wooden picture frame, detailing my freelance service as a licensed insurance assistant. I was hoping he would allow me to place it in the reception area where frequenting agents would notice it. Cayce was the guy everyone loved. He was well connected, and he knew where the opportunities were and who was looking for people. After I shared a few highlights from my trip, I confessed that I needed help finding some temporary work to replenish my bank account.

"Hey, our marketing assistant just left and we could use someone with your skills." Cayce grinned, satisfied that he had a solution for me.

"I can't," I said a little too quickly. "I appreciate the offer, I really do, but I don't think I can come back, not after all that happened."

"It's temporary and it's under my direction, not Vancouver. It would get you back in the scene and front and center to opening opportunities." He smiled, knowingly.

"OK, I'll think about it."

Having to make all these decisions was strenuous and I wasn't used to it anymore. The biggest decision of a day in Spain was which beach to go to.

A week later I walked through the doors of my old office. There was a new girl at reception and at every other desk. Again, I felt like a stranger stepping into the familiar. Cayce greeted me with open arms and walked me to a cubicle in the hallway.

"Ironic," I said, "this is just like the one I started in seven years ago," and we both laughed.

As I sat down to go through the employment form, I felt a churning in my stomach. *How had I come full circle?* Suddenly, sweat formed on my temples and the familiar heat of panic crept up my spine to the back of my neck.

"Welcome," a thick tongued man said from behind me.

"Thank you," I squeezed out a smile. *Didn't they know I was part of this team long before they were here?* I thought grimly. *Who is he to welcome me?* I was cranky and unkind, and I hoped my face didn't betray my emotions.

After the initial week, going to the office got easier. A welcome side effect of being employed was that I was eligible for credit and with my mom's co-signature, I got a smoking deal on a VW Rabbit trade-in. I was slowly gaining back my independence and freedom of mobility.

"Hey, listen Yvonne, I just got off the phone with Pamela at Assumption Life and she's looking for a business development person for the prairies. Isn't that great? You should give her a call right away." Cayce was a little out of breath with excitement.

I had met Pamela a couple of times in Vancouver and was struck by her presence and demeanor. A successful, well-spoken businesswoman who knew what she wanted and where she stood in life. *Another strong female leader I could learn from,* I thought. "I'll call her right away, thank you!" I stared at the piece of paper with Cayce's scribbles and hesitated. *What about my dream of opening the Wellness Temple? What if this job takes too much of my time?* I worried. I had watched many business development people over the years burnout from the pressure of meeting sales targets and practically living in airports. *Was I selling out for the sake of money? What if all the insights and dreams I discovered about myself in Europe didn't make it on the plane back to Canada with me?* I decided to wait and talk it out with my parents. After all, they knew me better than anyone and would tell me if I was stepping into a trap.

After dinner that night Dad and I settled onto the couch where Mom joined us with hot tea.

"You need money to survive here. You can't live from hand to mouth forever," my dad said. "The Ayurveda course you want to take is expensive and this opportunity will pay more than a Starbucks barista job, so you'll get ahead faster."

I understood the rationale, but I was concerned. *What if I got lost in the corporate world all over again?* Once I was used to a certain amount coming in every month, would I find the courage again to leave that security behind and follow my dream?

"What do you think, Mom?" I knew she would be less logical and more understanding of the emotional dilemma I faced.

"Yvonne, Dad is right. You can't live in our basement forever. You want to live your own life and that requires money," she said warmly. "You can take the courses on the side so that you don't lose sight of your dreams, and when you're ready, you can pursue that temple idea."

When I got to the office the next morning, I pulled out the piece of paper Cayce had given me and dialed the number. Pamela and the chief marketing officer flew in the following week to meet with me and sign the contract. It was a unique and enticing consulting contract at a competitive rate with performance incentives. The best of both worlds. The security that employment positions offer with the freedom to work on my own terms and in my own time. No one was looking over my shoulder, but Pamela was always there for mentorship. She helped me substantially cut my learning curve as a regional business development manager. We also agreed to focus on establishing one region at a time which minimized my travel, at least for the first year. I won't forget the day I received my first check. Not only was it more than I had ever made before but it was as promised and on time. With my professional life back on track, I intended to make a home for myself once again.

By spring, I had found a cute, one bedroom condo in an upscale downtown neighbourhood. The building featured floor to ceiling windows, a fireplace, and a bright den that served as my office overlooking the Bow River and connected to my bedroom with sliding glass doors. On weekends I'd browse the local buy sell boards where I found a gorgeous mahogany dining room set and a khaki colored L-shaped microsuede sectional. I was becoming the independent, professional woman I had always aspired to. I could take care of myself and wasn't controlled by anyone or anything. Or so I thought.

After another "day at the office," I opened the fridge and stared at the jar of pickled beets, a carton of milk, and leftover pizza. *I guess I'll have Pho again*, I thought. The tiny Vietnamese restaurant across the street had only a couple of tables, catering mostly to busy career folks who grabbed their takeout on the way home from the office. Some days I ate there just to be around people but most days I took it back to my nest, and carefully balancing the giant bowl of hot broth in my lap as I sat on the couch, watched the news with a glass of red wine. When I was finished, I'd immediately take the empty Styrofoam container to the downstairs trash can to avoid stinking up my apartment like a Vietnamese noodle house. On my way back, I'd detour into the courtyard where I sat, cigarette in hand, pondering my lonely existence.

Smoking had always been my social crutch, and now it was my companion. The alcohol-nicotine elixir was effectively numbing my feelings and made facing my being alone again on a Friday night a little less painful. My friends were busy tending to their new roles as spouses and mothers. If they had a night to spare, the last thing they wanted to do with it was to drive downtown or go to a noisy club. Being the third wheel or the spinster auntie at their family tables was OK sometimes, but as my longing for a different life grew, I withdrew more into my cave. I yearned for Spain, Italy, and Greece and regularly browsed the internet for the next flight out of Calgary.

***

The sun was already high in the sky when I rolled over and felt something sharp poke me in the ribs. I sat up startled. *What day is it?* I pulled the covers off me, too hot to assemble my thoughts. It

82 | Yvonne Winkler

was Saturday and I had been up late into the night again, clicking through online dating sites looking for love.

Understandably, most of my friends had learned to live without me, after being gone for over three years. Without a team in Calgary, the consulting work I was doing for Pamela and our client in Eastern Canada left me isolated professionally as well. Occasionally, I saw other company representatives at conferences, or we crossed paths in airport terminals, but that was the extent of any contact I had with my peers.

The long road trips through the flat prairies were less dull thanks to audiobooks, and with each trip to Edmonton and back I upgraded my professional skills. I learned how, according to Seth Godin, to be a purple cow and transform my work by being remarkable. I got curious about how Tim Ferriss suggested I could do more of what I love by working only four hours a week, and I developed Kim and Mauborgne's blue ocean strategy to maximize our competitive edge in a sea of insurance products. My confidence grew and so did my income, cashing in bonus checks every month which I reinvested into more professional and personal development. Steadily, I got back on track with the active and corporate lifestyle I had left in Vancouver. On weekends, I spent ninety minutes in back bending yoga poses followed by another ninety minutes karma cleaning the studio while filling my ears with more information about the brain-body connection from Dr. Wayne Dyer, Louise Hay, and Eckard Tolle. In short, I kept myself busy. The more books I consumed, the hungrier I became for knowledge—an effective distraction from my ever-growing feelings of abandon while helping me gain insights into a new way of life. They called it mindfulness and conscious living.

One of my last single friends, Carla, loved going backpacking in the backcountry of the Rockies and was willing to teach me how to survive in the wilderness. I observed how to pack the lightest backpack and have all the supplies one needs for an excursion in the woods, how to prepare a deep-dish pizza over a single burner camping stove, and how to connect with nature and embrace all of it, including overcoming the natural obstacles and my debilitating fear of bears.

As I drove up the gravel driveway in my black VW Rabbit, Hector, a solid one hundred-pound, white American bulldog with light brown patches greeted me with a wagging tail. He loved seeing me because it meant we were going on an adventure. We booked off the long weekend in July, and on Thursday night we spread out the tent, air mattresses, food, and changes of clothes on the floor of the one-bedroom suite she rented on her friend's horse ranch. The next morning, we loaded up her blue Jeep Wrangler with our packs, more water, and Hector, and hit the road as the sun rose in the rearview mirror.

"How far are we hiking today?" I bit into an apple and chased it with a handful of walnuts.

"Oh, I don't know, maybe twenty kilometers. We'll see how we feel at lunch break and decide." Without taking her eyes of the road, she continued, "The whole hike is fortyish kilometers. If we can get to the summit today, then we can spend all day tomorrow exploring around the lake and hike out Sunday."

Despite her small frame, she was as strong and conditioned as an athlete. Her country girl roots made her tough as nails, and she had hiked the Rockies solo most of her life. Now a business development career woman like me, Carla always had her rollerblades and hockey gear in the back of her car, ready for a quick twenty-kilometer skate between meetings, or to jump in as an extra for any league short a center player. I suspected that intense exercise was how she coped with her demons, given that not even a ginormous hematoma from being bucked off a horse slowed her down enough for me to catch up. Luckily, I had Hector. While he always led the way, he insisted on waiting for me if he sensed that I had fallen behind or was completely out of sight.

"Are you taking pictures of flowers again?" Carla asked, hiding her exasperation behind a big smile.

"It's not just about the destination, you know! You have to stop and smell the flowers along the way. What's your hurry anyway?" I didn't want to be the weak link, but at this pace, I didn't know how far I would be able to go with thirty pounds on my back.

"I just want to get there before we lose daylight."

"Oh c'mon! I didn't realize it was a race." I took a gulp from my water bottle, buying myself another quick moment to catch my breath. "Wait a minute," I squinted into the sun to get a look at her face, "are you trying to catch up to those guys from the parking lot?"

"No," she said a little too fast and turned to mosey onward and upward without having broken a sweat.

"You're totally racing up this mountain because you can't stand that they passed us!" I yelled out laughing, trying to keep up. "Well, too bad, I want to enjoy this and not grinch with pain every step of the way. I'm not as fit as you. I can't keep this pace for twenty kilometers."

Carla stopped, looked up the path and back at me. I guessed she probably didn't want to listen to me whine. She hated whining.

"Okay," she sighed. "You set the pace then."

I strode past her with a gratified smile, and we picked up another song to announce our presence to the bears.

"I think I'm ready to start dating again," Carla said out of nowhere.

"What?!" I turned around nearly stumbling off the edge of the path. Carla had been single for at least the two years I had known her, with many wooers, who she regularly turned down gently and firmly. This acknowledgment was a big deal. When we reached the summit a short while later, we both fell over onto the packs securely strapped to our backs, out of breath.

"Maybe spending the long weekend in the backcountry with you isn't the best way to meet a man," she said giggling. "We should have gone dancing."

Freedom Seeker | 85

Calgary was in the mid-thirties, so we were surprised by the three feet of snow up at Three Isle Lake. I could never sleep well in tents and that night Hector got so cold he squeezed between the two of us with much of his body weight against my hip. The two and a half inches of padding my Therm-A-Rest sleeping pad offered was barely enough comfort for my aching legs and back, and then there was the matter of the bears. I spent all night listening for any rustling in the bushes that would alert me before the grizzly had a chance to tear its long claws into our ultra-lightweight tent and maul us to death. I was tremendously relieved that we packed our gear and hit the trail early the next day so we could make it back in time to go out and find us some men.

That was my last backpacking weekend with Carla. She met the love of her life three days later on a dating site I helped her with after we struck out in Calgary's bar scene. As it often happens, the new relationship consumed most of Carla's free time and given how long she'd been on her own, I guessed she had a lot of making up to do. She invited me to hang out with them, hot tubbing and other mixers, but I felt like the third wheel once again and declined after a few.

We tried to maintain our friendship as best as we could. We squeezed in lunch or coffee whenever we found room in our already packed work schedules. I was happy for her, but I missed our quiet time and long, deep conversations in the mountains. Everyone I knew was now in a relationship, and I became what I dubbed the 'weekday friend'. No one had time for me on weekends anymore. Weekends were reserved for the family.

With nobody left to hang out with, I redirected all my attention to keeping alive the vision I saw for myself on the beach of Tarifa, starting by incorporating my company, Lotus Consulting Inc. The name was inspired by the flower's symbolism and the transformative experiences the wellness temple was going to offer to my customers. It also appropriately represented my current business development work that was effectively funding the research and development of the actual space. To me, it was the perfect name because the lotus strongly anchors her roots in the life-sustaining element of water. The stem and the pointy bulb grow through darkness to reach the light, emerging above the water's surface where it opens its big, multi-petaled blossom and radiates into the world. At night, the lotus closes to control the plant's inner circulation of water, a metaphor I began to use for self-care and preservation.

One day, sitting hunched over at my desk with the space heater blowing hot air on my legs, I was sifting through hundreds of work emails when I came across one from Inc. magazine advertising a series of webinars about how to write a business plan to get funding. The investment was US$300. I bit my lip and with a jittery hand typed in my credit card details without hesitation. This is what I needed, hand-holding from the experts through a step-by-step plan. Hopeful and excited to be moving closer to my dream, I dove into the homework from the first week, which entailed researching businesses that were like the one I wanted to launch.

I gulped when I discovered how many millions of dollars I needed to create this magical place of pools and saunas, tranquility rooms and medicine tents. *How was I going to raise that kind of capital? Who would invest in me?* I had no background in the spa or wellness industry and getting that Ayurveda training under my belt would take months and require attending a school in California or traveling to India, adding more costs. And even if I found someone who'd give me the time of day and listened, I didn't know how to pitch the idea. The deeper I looked into the seismic crater this temple idea left, the bigger the leap of making the dream come true. With a deep sigh, I put the folder labeled Lotus Wellness Temple back into the filing cabinet and slammed the door shut. I needed a smoke, damn it!

I had managed to refrain from smoking here and there for a little bit but nothing lasted more than a few weeks. I put on my coat, patted my pockets for lighter and keys, and stepped into the elevator to go out in front of the building. The cool, crisp air of fall bit my face and I quickly pulled up my hood before the next dust clouds whipped around the corner. It also served as a bit of a disguise. I'm not entirely sure who I was hiding from, but just in case someone I knew in a professional capacity happened to walk by my building, I didn't want them to see me smoking. Ashamed of not being able to get this under control after returning from Europe, I stood under the large overhang of the apartment building entrance, sheltered from the elements, and watched the blue cloud puffs dissipate into the air. *Who was I kidding?* My Friday night companions were a bottle of Shiraz and the background noise from the television. *How am I supposed to run a wellness temple when I'm drowning my loneliness in wine and cigarettes?* I thought. I needed a new plan, to meet new people and stop feeling sorry for myself. A text alert startled me back to reality. It was Jennifer who wanted to see if I had time for a coffee.

***

Between her new job demands and recent endeavour into a personal development program, we hadn't seen each other in a few weeks.

"Oh my gosh. It was so amazing! I feel like a brand-new woman," Jennifer gushed as we sat down at the window table in our favorite little café on Sunday afternoon. She gave me the highlighted version of her transformative experience. She was indeed glowing. This event, unlike other rah-rah seminars that usually tapered off after a few days, had really made an impact on her at a deeper level.

"Yvonne, I want you to experience it too. They have another one starting next month and I have a ticket for you," she nearly jumped out of her chair with excitement.

88 | Yvonne Winkler

"Wow," I stuttered. "Thank you. I don't know what to say. Wow."

"I want you to feel the way I do right now and have everything you dreamed of on that beach in Spain."

Since my weekends were mostly clear, I accepted her generous gift. *Maybe I'll gain the confidence I need to approach multimillion-dollar investors,* I thought dialing the number she'd given me to reserve my seat. *Maybe it's a cult. Oh God, what if they brainwash me and I can't get out?* I quickly dismissed the thought, Jennifer seemed fine. I had nothing to lose and so much to gain.

The first day of the retreat, I arrived early and took a seat in the middle of the last row of chairs, close to the exit, just in case shit got weird and I needed to escape. I wiggled nervously in my chair, hugged my body and buried my chin into my scarf. I had never enjoyed group learning or team events. A scene from my childhood flashed before my eyes, standing up in front of class with all the other kids and my teacher laughing hysterically when I got the question "What does the porky pine do in the winter?" wrong. I much preferred the solitude of a book or one-on-one coaching.

"Hi, I'm Tina," a tall woman with curly, unruly hair fell into the chair next to me. "I hear we're not allowed to take notes, wonder what that's all about?" Her voice matched my suspicious sentiment.

"Yvonne," I said, and we shook hands. "Yeah, my friend mentioned that the facilitator believes it takes away from the experience. We'll retain what we're supposed to, I guess." I folded my arms back in front of my chest and shrugged. At that moment loud, upbeat music interrupted our conversation. It was the cue to get settled and ready.

So much information was packed into three short days, with my takeaway being what a personal disservice it is to blindly follow rules and the danger of saying 'I know'. These two frameworks shattered everything I had learned from my communist roots, where my survival depended on never questioning authority and always doing whatever the signs said. I had never confronted this mindset or saw how it spilled over into aspects of my adult life. Until that weekend, I would detour the extra block to the pedestrian cross walk instead of jaywalking, and I never ever used the handicapped bathroom. *After all those years of waiting cross-legged in the bathroom line ups, nearly peeing my pants, when I could have used the perfectly empty extra wide stall.* I shook my head and smiled incredulously at the thought.

At home that night I picked up my favorite gel pen and continued to capture whatever I could in my journal, afraid that by tomorrow it would all be gone. *Pretending I knew it all and unquestioningly obeying whatever was put in front of me had kept me safe, but did it still serve me now that I lived in a free world where diversity, self-actualization and affluence can be possible?* I scribbled down hastily. Could vulnerability and curiosity be the key to making my dreams come to life? The exact opposite was instilled into me in communist Germany. I learned distrust and self-reliance were the only safe route to survival. This was the moment I realized I was in complete control over my own choices.

It was indeed a life altering event that led me to promptly sign up for three more. I felt like I found my community. Learning, growing, and meeting new people. This was the answer to my dull, lonely weekends. I desperately wanted to recreate the freedom I felt on the beach in Tarifa and fix whatever was wrong with me and stopped me from being highly successful. Personal development became my new obsession. Law of Attraction seminars, confidence building weekends, constellation healing, astrology readings, and belief repatterning workshops replaced my loneliness, drinking wine, and smoking cigarettes. I bought crystals and a salt lamp and smudged every corner of my condo. I immersed myself completely in new healing modalities because the traditional prescriptions weren't strong enough anymore.

One cold winter's evening two of my new, single friends invited me to dare the impossible—walking on fire. It was incomprehensible to me how I would do this without burning the soles of my feet, but it was just the kind of fix that filled my cravings bucket. I parked my car in front of the 1990s, crème colored bungalow in a southeast neighbourhood, and took a deep breath. *Was I ready for this?* I had done some wild things these last twelve months, pushing the limits of my comfort zone. I jumped out of a perfectly good plane for Carla's fortieth birthday, I did the flying trapeze in a trust exercise, not to mention traveling alone through foreign lands. *Yeah, I was ready.*

We were ushered into a dimly lit basement room with blankets and cushions on the floor. Before I joined the others in lotus pose, I went to the bathroom one more time to ensure nothing was going to disrupt this experience. The facilitator began to explain the course of events while I looked around the room to familiarize myself with ten strangers. Meanwhile outside in the snow covered backyard, a wood fire burned to a coal bed at about six hundred and fifty degrees Celsius. We were guided through an extensive mental preparation ritual that transformed our fears and put us in a state of consciousness where we could confidently step onto a blessed carpet of smoldering red embers without feeling pain or getting burned.

My heart was beating louder than the drum the facilitator began striking the moment we began the evening. We cheered loudly as each person bravely crossed the embers. Now it was my turn. I looked up to the sky, folded my hands and whispered a prayer. And like I was jumping into a deep pool, I stepped onto the orange and red flickering coals. My friends greeted me with a big hug at the other end, which brought me back from that special transcendental state my mind had achieved. That night I felt superhuman, an induced sense of euphoric achievement.

"If I can walk on fire, I can do anything!" I announced to the two friends with a grin that reached from ear to ear.

The experience was truly transformative, and everything I had learned so far about the mind-body connection was shown to me during those four hours. I began to see what was possible when I connected to my heart and became very interested in learning how I could do that whenever I wanted.

During my exploration I came across Amatsu, an ancient Japanese healing therapy. Dr. Christian Perez was an extremely fit and healthy-looking man with an almost militant attitude that was very confusing given the eastern medicinal nature of his craft. His Canmore office, a beautiful wood framed loft with bamboo ornaments and Buddhist prayer flags, was a good hour drive from my place. In addition to a rigorous and, for me, often painful soft tissue treatment, he also taught Zen meditation for beginners on Wednesday nights from six until eight. That's where I met Mini Gayzer. Mini, an account executive at a downtown engineering firm was first generation Canadian from India. We hit it off immediately, bonding over a shared dismay about our immigrant fathers who came to Canada for a better life yet insisted on maintaining our cultural traditions. We carpooled with two other students from Calgary, which made the two-hour roundtrip easily tolerable.

"Zen is a meditation technique rooted in Buddhism," Dr. Perez began once we were all settled. Because I couldn't twist my legs into a perfect lotus pose to keep my spine straight and hips comfortable for the two-hour class, I had bought a meditation bench which rested under my bum with my knees on the floor. "This style is different from other types of meditation in that practitioners keep their eyes semi open, focusing on a spot about one foot in front of you," he continued.

The combination of the time of day, the stillness after a full day of meetings and the ambiance of the small room, lit by the moon, inevitably made me want to close my eyes and fall asleep.

"Isn't sleep the ultimate form of meditation?" I blurted out in the car on our way home.

"The Zen technique is about regulating attention," Oliver replied studiously.

Typical Yvonne, I always bite off more than I can chew, I thought. I did the same in Spain. Learning Spanish in Andalusia was like learning English in Newfoundland. Beginning a meditation practice with 'thinking about not thinking' was like learning to ski on a black diamond run.

We traveled to Canmore together for the duration of the program. My posture improved, my nervous system calmed down, and I loved the depth and breadth of our conversations. Whether in class or in the car, I gained valuable insights about how to take care of myself differently by connecting to my body. At the same time, my consulting business had also gained a new level of momentum. The company I represented had launched an industry changing app which required three times the work hours than the previous two years. I gave up my karma cleaning gig at the hot yoga studio and sold my ninety-gallon fish tank that I rarely got to enjoy because I was never home anymore. Cleaning it took up most of my Saturdays, which was time I needed to catch up on my personal life. Eventually, I was missing more and more Wednesday nights, either because I was out of town or on a plane, so when the time came to renew for another season, I made the difficult decision to leave that too.

A year full of personal growth and development called for a big ending and solid start to a new beginning. I rented a bed-and-breakfast in Golden, British Columbia, and invited anyone who wanted to say goodbye to the year dashing down slopes, drinking hot toddies and waking up surrounded by white mountains on the first day of the new year.

"Is it OK if I bring a friend?" Jeremy asked through the speakerphone in his truck.

"I guess so, considering that you're almost in Golden now. It would be rude to say no!" I replied, annoyed. *Didn't he know that*

I invited him to see if we could rekindle the flame? Why was he bringing a friend?

"Tell him he can come if he brings a bottle of Gibsons," Jennifer piped in from across the table.

"I think we have enough booze for the weekend!" I glanced at the kitchen counter where we had stocked two one-and-a-half liter bottles of Grey Goose Vodka, a twenty-sixer of Crown Royal whisky, Scotch, and a case of red wine for our three-day stay.

"Are you almost here?" I asked turning my attention back to Jeremy.

"Just about, we'll just stop and grab some beers and head over. See you shortly." He hung up and I turned to the others who were playing another game of Go Fish.

It wasn't quite dark yet when they walked through the patio doors into the backyard, holding up a bottle of Fireball. Like it had been rehearsed, we all looked up, raised our glasses and yelled in unison, "Jeremy!" I was delighted to see him again.

I was excited to see if sparks would fly and possibly begin this upcoming year with a partner. We originally met ten years ago at my first job in the insurance business. He was in a serious relationship at the time, and I was engaged. Then we accidently crossed paths in Vancouver just before I left for my Europe trip, and although I was thrilled to see him, I wasn't interested in getting my heart entangled in a romance out of fear I would cut my trip short, or worse, not go at all. This was a great opportunity to see if we could pick up from where we left off.

"Hey Yvonne, this is my friend Dennis," Jeremy stepped aside ever so slightly, and the most intense blue eyes met mine.

"Thanks for having me," Dennis stuck out his hand, "Here's the whisky you ordered."

"Jennifer, your Gibson has arrived. Everybody, meet Dennis," I yelled over the music and across the fire pit, and turning back I said "Dennis, meet everybody."

94 | Yvonne Winkler

In short order I uncovered that he was Jeremy's boss and that he was only staying with us for one night. He had to be back in Edmonton to meet his girlfriend and attend a steak and lobster New Year's Eve event. *Why are the good ones always taken?* I mused as I took another sip of my vodka soda. Jeremy and I discovered quickly over the course of a couple of days that the flame was gone. Something wasn't right, like he wasn't available. I had really hoped I wouldn't have to begin another year alone, but I also wasn't going to settle for anything short of magical ever again. My biological clock was ticking, and I wasn't interested in another broken heart. I wanted to meet someone I could build a home with, settle down and grow old with. So, Jeremy and I parted ways and agreed to keep our relationship in the friend zone.

In February, standing in the NEXUS line at the Calgary Airport, I glanced at the person behind me as I was speedily unpacking my laptop and little bag of liquids and gels.

"Woah, what are you doing here?" I said perplexed, trying to recall his name.

"Oh hey, I'm off to Vancouver," he said.

"How funny is that? Me too. Have you gone skiing since New Year's?" I asked trying to buy time to think of his name.

"No, I've been busy with work. You?"

"Me neither. I don't really have anyone to go with," I said, moving through security.

"I can call you next time I go out, if you want."

"That would be awesome!"

We were both waiting to collect our belongings, he was putting his belt back on, and I was stowing away my liquids. *Come on Yvonne, think! What was his name? I should have asked him right at the beginning, now it's too awkward,* I chastised myself.

"I guess I should give you my number." I was hoping I didn't sound too desperate or inappropriate. He pulled out his phone and made note of my number.

"Are you going to terminal B?" I asked.

"No, I'm two hours early, I was going to head into the lounge and work from there."

"Oh, OK. Well good to run into you. Safe travels. And call me when you're hitting the slopes next." I tried to sound cool and relaxed.

"Will do."

As soon as he was gone, I texted Jennifer.

"Hey Jen, remember that guy Jeremy brought to Golden? What was his name?"

"Dennis," she replied promptly.

***

He called me from a beach in Cuba a couple of weeks later to ask me out when he got back in town.

"Umm, don't you have a girlfriend?" I asked surprised.

"We broke it off. It was coming for a while but spending the holidays with her family confirmed it for me."

I was reluctant to say yes because I didn't want to be the rebound girl.

"I don't know, Den, can I call you Den? Dennis reminds me so much of Dennis the Menace, and that really doesn't help your situation." I giggled.

"I've been called worse," he chuckled. "What's your hesitation?"

"Well, I'm looking for a life partner, not a fling. You've just ended things with your last relationship. Don't you want some time to regroup?"

"It's just a date not a marriage proposal," he laughed.

"Of course," I said, feeling rather silly for getting ahead of myself.

We met for drinks and enjoyed each other's company so much that we began hanging out at least once a week. We even managed to get a couple of ski trips in before the end of the season.

He was an excellent skier and showed me new tricks for overcoming my fears on the steep stuff. We talked about everything, and I was struck by his maturity, wisdom, and depth. I discovered that Den was married before and that he had an adult daughter who worked with him in his business. He loved blowing off steam from the day-to-day stress of running his small company by getting out on his Harley, cat skiing and traveling to sunny destinations. He was also thirteen years older than me which didn't bother me in a friend, but as a life partner I couldn't downplay the implications this had for me. He was in a very different place in his life than I was. When he looked into the future, he saw less work and more travel adventures. I had just begun to become the woman I wanted to be, free and independent, working on my dream. I was just getting started on life, he was looking to slow down.

Then there was the small matter of not wanting any more children.

"I'm too old to start all that again," Den said as we were driving to Canmore for his shoulder surgery appointment. "I've been there, done that and don't really need to go back," he added apologetically.

"What about marriage? You've been there, done that too. Does that mean you never want to get married again?" I grabbed the steering wheel a little tighter and braced myself for his reply.

"I would marry you," he said without skipping a beat.

"Wow, how can you still believe in marriage after all that you've been through?"

"I've never stopped believing in the institution of marriage. It wasn't all bad. There were good times too."

As he got out of the car I wanted to jump out and go with him, or at least give him an extra long hug and tell him how much I liked him. He wasn't going into open-heart surgery, but my entire body wanted to embrace him. Instead I said, "Good luck! Call me when you're out of surgery. I want to know how it went," and drove away watching him in my rearview mirror as he walked into the hospital alone.

Den had every quality in a man I listed in my brown leather journal on the plane home from Europe. He was handsome, adventurous, curious, successful, and he made me laugh. When he looked at me with his piercing blue eyes my legs went to Jell-O, and my entire body felt like it was on fire. I wanted to be with him all the time. The decision before me was agonizing. I didn't have to decide if I wanted to marry him, I had to resolve if I could live with my choice to never experience motherhood.

We continued to remain friends, but holding that line became more difficult for us as the months passed. I asked everyone I knew and valued to give me input. Of course, it was impossible for anyone to tell me what to do, only I would know what the right choice was for me. But I wanted other perspectives. I chatted with women who were married to older men and inquired what that was like. I asked the few friends I knew who didn't have children the deeply personal question "Do you have any regrets?" with a great deal of sensitivity and compassion, of course.

Unpacking thirty-three years of conditioning that my path as a woman was to be a wife and bear children was pushing the limits of my personal development. On my quest for answers, I discovered that most of the weight had to do with external, social expectations and little with an internal longing to push a human out my vagina. It felt like motherhood was the ultimate form of substantiating my success and worth as a woman. I recalled a dialogue with two partners in a financial firm who were looking to recruit me to work for them as a junior partner/executive assistant, three years earlier.

"Don't you want to be a mother?" one of them asked, leaning back in his chair with his arms behind his head.

"Sure," I replied.

"Well then you better get your career on track! You won't be able to do both," he pointed out.

I left that lunch meeting doing the math in my head and mulling just how I could swing both to prove him wrong.

Growing up, I didn't play family often, but I proudly walked my brown-haired doll in a blue stroller to my grandma's house before replacing the maternal role-play game with playing teacher or riding our bicycles along the corn fields, pretending we were on motorbikes.

Now I was wondering if maybe I wasn't meant to be a mother in this lifetime. Maybe my soul was here to heal some things and be of service in another way. The expectation to deliver a grandchild for my parents, and the fear of missing out on experiencing the unconditional love that bonds a mother and her child, was agonizing. What if I made the wrong decision? I was already thirty-four and my childbearing years were closing in on me. I couldn't undo this.

6
BURN OUT

Calgary, Fall 2015

After months of torturous restraint of our growing love for each other, I couldn't take it anymore and jumped into our relationship with both feet. I reasoned that if things didn't work out between us, I could always adopt. It had become clear to me that I wanted a partner in life more than I wanted to be a mother. We moved in together, putting an end to our ridiculous travel schedules and living out of our suitcase for months on end. Two years later, on August 8, 2015, we tied the knot at a private mountainside acreage in Golden, BC, the place where we first met.

The leaves were already turning yellow and brown when we returned from our three-week honeymoon through the United Kingdom. We both had a few more work trips before the year was a wrap, and I was deep into planning and setting goals for the next year. I printed out a twelve-month calendar, and with a bright yellow pen, began to colour in all the long weekends and holidays. Den and I escaped our growing work demands, bursting inboxes, and family expectations by traveling to new places as often as time allowed. On top of the calendar I wrote my personal and professional goals in big bold letters. It read "Get an Audi" a purchase I immediately regretted because it felt wasteful when I realized that the reason I got it was because I wanted to attract more abundance by looking the part. There was something about getting comfortable with money and the realm I would need to play in to realize a ten-million-dollar wellness temple. Truth was, I was terrified to make that kind of money. I wanted it, but usually I downplayed my financial achievement because I didn't want to standout or be disliked by my friends. I had resisted success for years and kept my big dreams at bay as an excuse not to leave the comfort and security of my family's story, and the deeply entrenched belief that business owners make their life on the backs of others.

It was a Friday evening in November, the fireplace was on, candles were lit, and I was making popcorn for our movie night when my phone rang, and I saw a name I hadn't seen in a while on the display. I instantly knew something was wrong.

"Hi Meena," I said cautiously.

"Yvonne, I'm calling to let you know that our friend Mini passed away last Wednesday." Meena took a deep breath swallowing her tears and continued, "I thought you might want to attend the funeral. I know she would have wanted you to be there." I braced myself on the counter, and not knowing how to respond to this news bomb I mumbled,

"I can't believe it. It's all so sudden. Did she know? I had just texted her a few weeks ago to see if she wanted to go for a yoga class." I reasoned.

"Yes, her cancer had come back, and I believe she knew for a while but kept it from all of us for as long as she could. I found out when I went to visit her and was redirected to the hospital. She didn't want any visitors. You know the fighter she was, but the cancer was already too far spread." She proceeded to give me the funeral information as I fumbled with a pen and paper to jot it down. "The meditation group and Dr. P are all going to be at the funeral, we'll hold a seat for you, OK?"

"Okay," I uttered and hung up before I was too choked up to speak. Tears began streaming down my face and my heart ached for our loss.

Den jumped up from the couch. "What happened?"

I pressed my face into his shoulder and sobbed. "Mini's cancer came back, and I knew something was up when I read her blog last month." I pulled away, using the back of my hand to wipe the tears. "She wrote about how overwhelmed she was, so I reached out but I got a super short text reply. Totally unlike her." I could feel anger coming on. "I hadn't seen her since my stagette in August. And between the stress around our wedding and then our honeymoon, I lost touch with her. And now, now she's gone."

Mini was in her early- to mid-forties, only a few years older than me. She was born in New Delhi, raised in Calgary, and, much to her parents pleasing, pursued a career as an engineer in oil and gas. It was a very reputable position, and she worked many hours overtime to maintain her title and reputation with the firm. After I gave up the Zen meditation class due to my chaotic work schedule, we met occasionally for tea and yoga at a studio close to her house. We could talk for hours about juicing, taking care of our bodies, spending less time in front of a computer screen or with people who don't love us as we are. Over rooibos tea, she confessed that she regretted how much time she had given to her career instead of doing more of the things she wanted to do. I confided in her about my challenges around choosing the unconventional lifestyle of not having children, something she had also been through.

When I finally arrived at the funeral home the next morning, the ceremony had already started. Meena gave me the name but in my shock from the news, I hadn't paid much attention to the details and arrived just in time at the wrong location across town. Out of breath and patience, I quietly snuck in the back and leaned against the wall with a few other people who couldn't secure a seat either. From the front of the oval room, paneled in fabric and wood, I could hear a quiet whimpering. *This must be her mom*, I thought. Her husband, Russ, whom I had met only the once at our holiday party, stepped up to the podium.

"Please close the casket," he requested calmly. When nobody moved he added, "At least while I'm saying goodbye," and

102 | Yvonne Winkler

gestured to the funeral attendant. "That's what Mini wanted," he whispered.

Her parents even disregarded her final wish, I thought. Russ was a white man with blond, shoulder length hair who taught guitar lessons and played in a band. Mini and I seldom talked about him and her parents' relationship, but I had gotten the distinct impression that it wasn't easy for them to accept him, a non-Indian. The tension between them was certainly palpable now. I looked up at Mini's picture on the white screen, smiling her cheeky smile at all of us. *Damn right, that is what she would have wanted,* I thought. With the casket closed, Russ continued to give a heartfelt eulogy for his wife. He honored her every wish and reminded us what Mini stood for and how hard she fought to reclaim her life after her first diagnosis, and not just from cancer!

Russ stepped down from the podium, and as the funeral attendant opened the casket again, my mind wandered. It had been five years since I had broken the shackles that tied me to a desk, and what had I done with all those good intentions? I'd worked myself right back into the prison of debt and job security. I was working more hours than ever before and spending more time on the road and in airports than seeing my friends and family. I touched the oval pendant, the size of a small thumbprint, dangling from a delicate silver chain and recited the words stamped into the silver metal "Dwell in Possibility". It was her wedding gift to me, and when she put it around my neck she said, "Remember to choose your own freedom!"

Row by row, people stood up and got in line to express their condolences to Mini's family and to say their final goodbye. My legs were like lead. Immobilized by the weight of anger and grief, I moved slowly with the other mourners. *Why did I not try harder to stay in touch?* I wondered. I was replaying the scene from my stagette when she gave me the necklace, trying to see if there had been any clues I didn't see because I was too wrapped up in my own life. Standing over her body, I barely recognized my friend.

Mini's skin was as smooth as porcelain, her face plump and grey from the embalming fluid. She was dressed in a white blouse with a lace doll collar, something I was certain did not come from her closet. I felt like I had let her down.

***

After Mini's death, something had shifted inside of me. I kept moving about my usual days at work, but her passing woke up the wild in me again, the same incitement I felt in Vancouver and on the beach in Tarifa. I opened the filing cabinet and pulled out the thick folder labeled "Lotus Wellness Temple". There were scribbles on various pieces of paper. One contained a complete sketch of the heating and cooling waterfall system I had designed for the center of the building. Clippings of ideas and photographs captured in different hotel spas where I had lingered in the serenity of quiet piano music on long business travels. There was the letter I wrote to the owners of a nearby resort after they lost their world-class golf course to a historical flood. My idea was to engage them into a collaboration with me that included an outstanding wellness retreat with hydrotherapy pools, saunas, and meditation space. The location was perfect. A hotel and conference center nestled into the Rockies, a short one-hour drive from metro Calgary. Unfortunately, I was too late. They were already in final conversations with a new developer who had the exact same idea, along with experience and capital.

As another roadshow circuit and the first anniversary of Mini's passing was looming, I grumbled in disappointment to myself as I dragged the heavy box containing all my tradeshow supplies from the far back corner in our basement. The next three weeks were a string of hotels, airports, and long days standing in heels at my company's booth smiling and rehashing our unique selling proposition, followed by restless nights due to late dinners and too much red wine.

Pamela, who was now overseeing a small wholesale team of seven of us, peppered from British Columbia through to New Brunswick, joined us at as many conferences as she was physically able to. We used this time together to bond and make business plans for the following quarter. As we relaxed into the leather barrel chairs in our hotel lounge after a long day of shaking hands, we ordered our favorite, a bottle of pinot noir and a cheese tray. She looked nervous, which was completely out of character for an otherwise self-assured woman.

"I have some news, and I don't think you're going to like it." She took a sip of her wine. "They want all of us to sign employee agreements before the end of the year." She scanned my face for a sign. "You had always said that the day they make you an employee you would quit . . . I'm afraid that day has come."

I took a deep breath and formulated my response carefully. I didn't want to burn bridges with a knee jerk reaction.

"I kind of had a feeling that was the direction the new management was going. They're old school and didn't seem to get us and what you had built. So, I'm not surprised to hear that they want more manageable control over their business development team," I sliced some cheese for my cracker and continued, "Let me think about it, and I will give you my answer when we meet again in two weeks."

This was my jumping-off point. The fork in the road that I needed to change the path I was on, and, if handled properly, without the drama I had experienced when I left Vancouver. I wanted to be strategic and discuss this with Den before making a decision that affected my contribution to our family. Letting this contract go also meant relinquishing my financial autonomy, at least momentarily.

"You've been dragging your heels ever since Mini died," Den said when I shared the news with him. "You don't like being on the road, and when you're here, you're so exhausted from it all you have nothing left in the tank to do anything fun, let alone another business."

We chatted about the Wellness Temple every time we visited a spa or hot springs for a quick R and R injection to get us through another month of doing something that didn't light us up. But it was always obscure, like dreaming about what we'd do if we won the lottery. Now that I was at a crossroads, I wanted to dwell in the possibility of my vision and give it serious consideration. Being a successful business owner, I valued his thoughts because he understood all too well the sacrifices entrepreneurship required.

"Managing staff is like herding cats," he frowned. "And besides, how is this going to work when we want to travel after I sell the company? A physical business ties us back down."

"You've been talking about selling the company for as long as I've known you. I don't want to hold my breath until that day comes." I gave him a sideways look and we both sighed. "Besides, I still want to do something meaningful with my life. I'm too young to sit and knit."

He did have a point though! Did I really want to commit to a brick-and-mortar business that tied me down for at least a decade of managing other people, the bureaucracy, environmental impact studies, and sleepless nights over how to pay my staff? I watched my parents do it and nearly lose everything over it. Den had been trying to get off that hamster wheel for many years too. Maybe I just loved how I felt when I went to the spa, the stress of running one wouldn't be the same as relaxing in one, and if I created a successful consulting business that afforded us a lifestyle, we could visit lots of them.

Somewhat relieved that I had a viable excuse to close the file on the Wellness Temple, I reluctantly offered, "Maybe I can write a book and become a speaker. That way I can do something impactful and we get to travel all over the world!"

He smiled and nodded. "Sure, that would be great." I sensed his doubt. "If that would make you happy, then I support your decision. I just want to see you laugh again. I can cover us while you get Lotus Consulting Inc. off the ground."

My whole body relaxed instantly at the resolute sound of his commitment to my happiness over a paycheck. I jumped up and gave him a big hug. *Is this what married life is supposed to be like?* I wondered.

"I could coach women who are in male-dominated work situations and help them get clear about what they really want so that they can live a life they love," I said excitedly. "And maybe that includes an annual retreat at a beautiful, serene spa that I'm not the owner of."

I confidently handed Pamela my resignation letter over dinner a few weeks later and agreed to help her find and train my replacement.

***

The first three months of my solopreneurship consisted primarily of reclaiming some personal time. I took my mom on a special birthday vacation to Hawaii, helped Den's daughter with her household as she had just given birth to a baby girl, and I got my holiday greetings out early for once. I was optimistic and excited for the next year although I had no idea where to start. One evening, cozied up on the couch with the fireplace roaring and my third glass of wine, I scrolled through Facebook when a picture of a white sandy beach, and a blond woman standing in front of a lush, green mountain backdrop caught my eye. She called herself The Bikini Business Coach. I clicked the ad and a short hour later was convinced that I needed her secret technique to getting my coaching business off the ground and make a quarter of a million dollars in my first year of business. She promised she'd give me the tools and help me with my money mindset.

The idea of having someone fast-track my learning and teach me everything I needed to know about creating an online group coaching program was terrifyingly exhilarating. I couldn't wait to get started and see the dollars roll into my bank account. In my mind I was already accepting the speaking invitations for sold-out conferences sharing my entrepreneurial success story. If this worked out as promised, I wouldn't need to mooch off Den's income at all, I would have the means to support my parents' retirement, and I could show all the doubters that I did have what it takes. I finished the last drop of wine and went to bed. My entire future rested on the success of my new business, Yvonne Winkler the business coach on the go. I was going to give it everything I had and make Den, my parents, and Mini proud. *I'll have to pay for it in two installments*, I thought as I closed my eyes. *But if I take the money I've saved and whatever I've got left on my credit card, I can pay for this without involving Den.* I didn't even want to tell him, after all, I shouldn't have to ask him about financial decisions for my business.

For the next thirty days I didn't come up for air. I tirelessly worked through the material, pouring my heart into writing my hero's journey, learning the steps to a successful Facebook ad and creating colorful posts for my private group, The Freedom Seeker Community, which came to me when I summarized forty years of my life into a five-minute highlight reel in one of the group Q&A sessions. I had been seeking freedom all my life, and what I wanted to create was a community in which we could have unfiltered conversations about what kept us trapped in unfulfilling situations. I sketched out the components of a ninety-day group program and overcame all kinds of fears that reared their ugly heads like going live, sharing my thoughts, and showing up as an expert in my field. My left shoulder ached so much I could barely reach for a coffee cup in the overhead cupboard. At night I was nervously grinding my teeth which left me with big canker sores on the inside of my cheeks and a very sore jaw. But I couldn't stop. I didn't want to lose one single thought, idea or moment to idling time like eating, exercise or sleep. Most of all, I didn't want to fail.

I set a launch date for my first online group coaching program for the beginning of spring because I needed to generate some income as quickly as possible. My Audi payments, along with other existing business expenses, not to mention all the new start-up investments, were draining my account faster than I could say, "Hello, my name is Yvonne." I was ready to attract hundreds of women into my business and start contributing to our monthly household expenses again, which Den had said not worry about until my business was generating regular income, but that was especially difficult for me.

I had been financially independent since graduating university fifteen years ago and proud of it. After a couple of breakups that left me financially holding the bag, I swore I would never again be so careless. I worked hard to climb out of that mess. Having money in my account gave me peace. Relinquishing a regular income with a mortgage, Audi payments, and a certain lifestyle left me feeling more vulnerable than I cared for. Suddenly, I was overcome with guilt over every dollar I spent, whether it was for milk or a cut and color at my reputable hair salon, because technically, it wasn't my money. Trusting that my husband truly meant the offer to take care of me was foreign and seemed impossible. No one would see my success for the struggle it was. People would think I had it all handed to me and, my absolute-worst nightmare, that I was with him for his wealth, that he was my sugar daddy. The mere thought that my love for him would stand in question devastated me.

In an effort to make it all right in my head and get a grip on my finances, I called my trusted friend Carla who, besides being my hiking buddy, had become a certified financial planner. She agreed to meet with me, and as we sat around a big boardroom table in her downtown office reviewing carefully compiled spreadsheets with numbers and graphs that made my eyes glaze over, she said,

"Yvonne, you have nothing to worry about. Looking at your combined wealth you can relax and really focus on creating the impact you want to make with your coaching business."

"I feel like I'm not deserving of any success this business might have because I didn't start it from nothing," I countered with a lump in my throat.

"What's that all about?" she asked perplexed. "You think that the only way you're deserving of success is if it was earned on crippling debt and debilitating hardship?"

Carla had a point. I had by now spent years studying mindset, the laws of the universe, and how to successfully attract abundance into my life, but my limiting beliefs about wealth were getting the best of me. It was like I was slapping the universe in the face with this lingering subconscious believe that money was bad and having it made me evil. Was it ancestral guilt? Was it the ten years of socialist conditioning that capitalism is inherently exploitative?

I left the meeting and headed home to get back to work. After this meeting I relaxed a bit about the money but only because I still thought that if I worked really, really hard and delivered a high quality, high value program, I wouldn't have to worry about this for too much longer and go back to being independent and free.

To attract a high-paying client, I needed a high-profile website and online presence. A friend was just launching her branding and marketing firm to collaborate with other businesswomen in order to rise together. This was also one of my values, and I became her first client. She built me a gorgeous website. Meanwhile, I sunk more dollars into promoting my upcoming program which I set up on an over-the-top online course platform. I wanted to look like a pro and have the quality of the experience match the dollars my clients were investing. To offset some of the start-up costs I learned all the ins and outs of writing, producing, and recording my own videos. After five hours of filming short twenty-minute lessons, I spent another six hours editing the footage to perfection.

"Dinner is ready," my Den poked his head into the slightly cracked door of my basement office.

"Ok, I'll be up in five," I said dismissively.

An hour later, I emerged to find him on the couch flipping through the evening TV programs. My dinner still sat on the counter where he had carefully plated the chicken and roasted potatoes.

"You'll have to reheat it," he said without looking up from the TV.

I could tell he was annoyed with me. He had been taking care of dinner for almost three months. Something I appreciated immensely because my desire to finish the program was so all-consuming, I had no capacity left for creating a nutritious, balanced meal.

"I'm concerned about you," he turned off the TV and turned his attention to me as I sat down with my reheated plate.

"Why? I'm almost done," I said breathing through a hot potato.

"You can't keep this up. This pace. You work night and day, and you don't come up for air. It's not sustainable." The strain on his face let me know that he was talking from experience. I wasn't going to let it consume me.

"The program launches in three days, and I have eight women who are expecting a high value experience. The next three months are not going to ease up but I'll take some time off during the summer before I launch the second group."

He shook his head in defeat and turned the TV back on. I put my empty plate in the dishwasher and made my way back to my desk to finish the editing.

I launched with only 3 percent of the total number of people I had anticipated which recovered approximately 80 percent of my initial investment into The Bikini Business Coach. It didn't recuperate any of my time or the website hosting and freelancer fees I had plunged into it. It certainly didn't earn me the C$250,000 that the Facebook ad promised. I was devastated and completely burned out.

I did everything the experts said. I invested heavily into my business and only asked of my clients what I was prepared to do for myself. I collaborated with other business owners and referred them to my clients. I kept an open mind to learning, and I recited my affirmations every day. I got visible, I made offers and defined my ideal audience. It seemed like it worked for everyone else but me. Daily updates on Instagram validated that. Something was clearly wrong with me. I needed to do better. I needed to become a better businessperson who can run a successful company.

My relationships, especially my marriage, were suffering hard under the weight of my aspirations. I didn't really have a social life, and I wasn't interested in talking about anything other than business or how I can be better at it. My friends stopped trying to connect with me, my family stopped asking how it was going, and Den buried himself in his own work. There were days I didn't speak until midafternoon when Den checked in to see what I wanted for dinner.

Being isolated, with no active feedback on my performance, only played further havoc with my confidence. I was alone again. I had no one to bounce ideas off. No one to give me real time feedback on what I was putting out there and no one I could talk about any of this with, except the coach I had hired to help me get through the emotional slog.

To ease the tension, slow down my thoughts, and forget about my failures as a businessowner, I began to drown my sorrows with more wine. At least that was my favorite. Some nights I would have vodka or gin or prosecco but the one that I could always rely on to give me the proper dose of satisfaction and narcosis was red wine. I had self-medicated with various drugs in the past but none came in as perfect a package as wine. It was the sophisticated escape with a deliciously intoxicating nose and perfectly acceptable in certain social circles, encouraged in most.

***

"I don't know what's wrong with me," I cried as I plopped onto the purple velvet couch in my coach's office. "Why am I so unlovable? Why does everyone pull away from me? Am I too intense?"

Tuesdays with Kerry, as I affectionately called them, always began with her greeting me at the top of the stairs, arms wide open, waiting to embrace me and bestow a smooth compliment to let me know she sees me. Anytime I was around her I felt unconditionally loved. Kerry was a wise sage who dressed in colorful leggings and big cozy sweaters. Her golden-red locks framed her smiling, round face and pink lips, and her light ballerina slippers completed her fairy godmother appearance. A friend recommended her to me as the best coach in town when I decided to branch out on my own, and within thirty minutes of our first session I knew she was exactly who my soul had been craving to meet.

Kerry loved the color purple, and every shade was represented in her small studio office. The magenta wingback chair in the corner right by the window was her throne, a faded lavender and mauve circular rug defined the coaching zone which was completed with two plum colored accent chairs. 3M Post-it Easel Pads with leftover scribbles from previous clients lined the opposite wall. She didn't care that her space might be too feminine for her downtown executive male clients. She knew who she was and who she wasn't, and I wanted to learn everything I could about the deep inner journey she was able to take people on in a nano second.

"What happened, angel?" Kerry sat in her chair across from me, her hands folded in her lap.

"I feel anxious," I sighed deeply. "I can't breathe. There's a tightness in my chest. My shoulder is so sore I can't lift it over my head, I even left the gym early yesterday, crying because I was in so much pain and frustration. And then there were the migraines over the weekend."

"Aw, Yvonne!" she said empathetically.

"I'm exhausted. I wake up at three a.m. almost every night now and can't get back to sleep. Most of all I'm tired of going over the same things with you, I feel like I'm letting you down and wasting my time," I poured my heart out, big, fat tears rolling down my dry cheeks. It was the only place I felt safe to do so.

"I can see that you're tired, angel. You do too much. You need rest."

"But it's September!" I said, appalled. "September is a new beginning. I took most of August off," I maintained in a defensive tone. "I need to get clients or close the doors to my business. I'm not ready for defeat. I can rest after it is up and running."

"Yvonne, how can you effectively be championing for women in your program when you're burned-out?"

Her words cut me sharply. Just like titles or an Audi didn't gain me more respect, or the cigarettes helped me to fit in, or creating an offer in six months, all on my own, wasn't going to shoot me up the success ladder. Over the next sixty minutes, we continued to unpack what got me to this dire place of burnout and depression, and I began to see that I was operating from an old paradigm, namely freedom equalled independence. But how was that viable or sustainable? I was exhausted from holding it all together and working endlessly to justify my lack of financial contribution to our life.

I knew that I couldn't ask my clients to trust, be authentic and true to their desires when I wasn't taking my own medicine. I was avoiding my feelings with overwork and playing it safe by talking and acting like I thought I needed to for the world to accept me. I needed to change the course of my life to match the work I had set out to do; engaging women in the possibility for freedom through bringing whatever work they wanted to do in the world into a viable, supportable, sustainable form.

"So, angel, what are you called to right now?" Kerry broke the silence.

"I need to retreat," I replied without hesitation.

"Retreat calls for a surrender," she pushed. "So, what are you surrendering?"

"I surrender that my vision has to look a certain way, in a certain order and within a certain time frame." I said definitively. "I was more focused on recovering my investment and making money than developing an offer that truly serves my audience. I let fear and not-enough-ness direct my action, not the wisdom of my heart."

"All of your unconscious fears give rise to the stories you tell yourself. Angel, you have the gift of the wound of fleeing oppression, of secret, of hiding, and of escaping. You can't go forward in life and create what it is you want to offer without feeling those feelings. The whole regime you were living under was about disabling you from taking initiative or stepping out of line and saying that the way we're doing this isn't working, let's try something else. Do you see? Your story is fundamental for humanity. We can't be about building businesses anymore. We must be about creating our future," Kerry declared firmly. "And we can't create our future when we're encumbered by our past. You might consider reframing yourself as a healer."

"What is it my work to heal?" I asked perplexed and genuinely at ease with the title.

"This place where humanity, women in particular, stop themselves because of unresolved childhood trauma. It's a block. And when we don't know what to do about it, we get busy doing anything else instead, like building businesses, marketing, sales."

I resonated with her words. She was describing me. I wanted to create a new path, new opportunities, new possibilities but kept getting trapped in my childhood trauma. What was worse was that we had been conditioned to process them in a psychological or

Freedom Seeker | 115

psychiatric way. We were traumatized. We needed to heal, not get fixed.

We didn't speak for a long minute. I was processing the implications of my discovery. Every word felt like a soothing comfort to my very raw heart. It had been months of my foot on the gas without checking if I was heading in the right direction, giving way to feeling like a giant failure when, after thousands of hours, my efforts weren't reflected in my results. By taking a moment and going inward I could see that my work in the world wasn't about trying to figure out how to cut expenses and get better at writing sales copy but rather to observe my own healing journey and look for direction there. Because when I healed my trauma, I could begin to hear other women's trauma and that's how I could help them bring their gifts into the world. Could the Lotus Wellness Temple be me? Was it possible that the serene space I had wanted to create for women was not a physical structure but a place within? Dwelling in possibility meant I needed to quiet my mind, not launch into action.

7

THE BREAK WE NEEDED

Over the next few months, I cut back on the amount of time I spent in my office. I decided to take Kerry's advice and not run another launch campaign. Instead I pondered how my lived experience could serve my clients in an authentic way. Looking around my workspace I noticed that everything, including the massive, steel framed corner desk, screamed productivity. There was nothing inspiring or nurturing about this space. With a paint roller, some gold accent pieces, and light colored woods, I created a more balanced, bright, and beautiful home office environment. A big meditation chair and a vintage floor lamp completed the space with an invitation to sit and meditate, journal, or read. The shelves beside my desk had frames of my favorite quotes and a picture of me smiling from ear to ear, holding my signed copy of *The Desire Map: A Guide to Creating Goals with Soul*, standing next to bestselling author Danielle LaPorte.

Curled up in my meditation chair with a blanket and my journal one gray afternoon, I replayed the recording of my last session with Kerry, looking for answers and wisdom I might not have heard while I was in her office.

"Do you think alcohol is impacting negatively on your life?" she probed carefully.

"No!" I shouted. "I mean, I'm not a binge drinker. I only have a glass or two a night. I don't need a drink when I wake up or anything, and I don't get blackout drunk," I had added quickly.

This wasn't the first time I had been confronted about alcohol and the role it played in my well-being. The question "Do you think alcohol is interfering with your life?" opened an old gash, and my thoughts drifted into a deep, dark corner of my mind where I had tucked away the memory of a night out in Vancouver, shortly after I had moved out of my ex's and into a small suite in trendy Yaletown. Ashamed about the events that lead me into the cold, white stone walled basement office of an addiction's councillor, I hadn't told anyone about this, not even Kerry who knew my soul by now.

It was a Saturday morning six years earlier. The sun broke through the cracks of my drawn blinds and tickled my face. As consciousness followed the invitation, I sat up promptly, not knowing what day or time it was. The sudden jerk made my head spin, and I rolled over the side of my Murphy bed and threw up. *Oh God, what happened?* I thought as I laid back and tried to retrieve the missing information. After a few minutes of letting my brain catch up to my body, I carefully got up to get some water and my Blackberry. There was a voicemail, and as I retrieved the message from my friend and co-worker, I began to clean up my mess. We had started with our usual happy hour Friday afternoon drinks. As, one by one, our coworkers disappeared home to their families, she and I decided to hit up a party hosted by a pretentious financial broker with a gold ring on every finger. I'd encountered his type many times before and thought nothing of it, as long as I didn't have to pay for the cocaine or vodka, I played along. Sometimes, that meant having to laugh off machoistic jokes or tolerate being objectified. That night, I didn't feel safe. I vaguely remember a man pulling his girlfriend behind him and ripping the host a new one about the obscene amount of drugs and alcohol present. That could have been my ex who would have dragged me out of this situation just like that, only I wouldn't have been in this situation if I hadn't just ended our relationship, and I wanted to prove that I was a strong, independent woman, completely in control of my life.

That night left me incapacitated for the rest of the weekend and proceeded with a depression that didn't ease up for several weeks and ultimately scared me into calling out for help. It felt like I'd had a lucky escape and some part of me knew the next time wouldn't turn out so well. So, I took advantage of the free integrated workplace health care benefit and saw a psychologist for the six weeks that were covered. Her words still rang in my ears.

"You don't have a drug problem," she said. "You have a drinking problem. Alcohol clouds your judgement and natural inhibition."

Following that incident, I experienced the benefits of sobriety for three months before I threw caution to the wind and cheers'd with my friends at the pub once again. I stayed far away from cocaine but not drinking alcohol was too antisocial for the people pleaser in me. People got weird and squirmy when I revealed that I just wanted soda, never mind uttering the word 'sobriety'.

I knew firsthand what not drinking would mean for my social life. Admitting I had a problem or that I'd been to A.A. would cause irreparable damage for a woman trying to climb the career ladder. So, I talked myself out of the problem. I had done it then, and I was fully prepared to defend that position again. Only, I couldn't fool Kerry. Our entire relationship was built on absolute and complete honesty. I also had a deeper awareness of myself now and couldn't deny that alcohol was killing my zest for life and my liver at an astonishing rate of one bottle of red a night.

"Damn it!" I cussed. Once again she bored right to the heart of the matter.

How did this happen? When did I lose the ability to stop at just one glass of wine? When did I begin to use it to self-medicate? My head felt like a major construction zone. Lots of moving parts, loud noise, and destruction to rebuild. I had done booze snoozes for short spurts of time, like sober October and dry January, but admitting that alcohol was influencing my life meant that I would have to stop, possibly forever. Given my relapse experience with smoking and several secret attempts to moderate my drinking, I knew it had to be all or nothing. Faced with yet another thing that was wrong with me, I stormed out of my office and straight to the liquor store, thinking *Why can't I be normal? Why does everything about me have to be so fucked up?*

The next morning, I woke up with a deep burning sensation behind my eyes. My body ached from tossing and turning because one minute I was too hot and the next I was too cold.

"You drank the whole bottle!" Den stood in the bedroom doorway his voice registering surprise. "Do you remember anything from last night?"

I tried to recall if something had happened but gave up. "No, what's wrong?"

"You called me an insensitive asshole. Again. I'm so tired of you picking fights with me when you drink." He turned away with disdain in his eyes.

"I'm sorry, babe. I didn't mean it."

I hated the person I became when the alcohol kicked in. Loudmouthed, opinionated, argumentative, and self-righteous. It was like the liquid gave me the courage to say what I needed to but discombobulated my words, leaving me unable to express what I meant in a clear and direct way. I also hated waking up every morning apologizing for what I had said or done the night before, even if I'd done nothing wrong per se, I felt guilty and ashamed.

"I will do better, babe. I promise. I'll take a break from drinking."

"Yeah right. Like that has ever lasted more than a day. Can't you just have one? Why do you always have to finish the bottle?" he asked, his jaw clenched.

"I don't know. Somewhere along the way the switch broke. I think I can handle one more and then it's lights out. I think there's something wrong with me," I said avoiding his gaze, my shoulders nearly touching the floor.

The mere thought of going to A.A. made my skin crawl. I'm not a drunk. I didn't belong there. I decided to try and quit on my own using some of the tools I had learned when I quit smoking.

If I can abstain from it long enough, eventually the urge will subside and the benefits will outweigh the pain, I thought. A.A. just made it more final than I was ready for.

***

Calgary, September 2018

I woke up with a jolt. Brushing my hair out of my face, I leaned over to my nightstand to check my phone for the time. Den and I had agreed that we'd keep our phones out of the bedroom, or at the very least on the dresser across the room, but on nights when he played hockey, I always brought it to my bedside just in case I had to dial 911. As the screen woke up, I saw that I had twelve missed calls. *Oh God! Den!* My heart raced and my skin crawled. I clicked on my husband's name in red letters, and he picked up on the first ring.

"Babe, I'm in an ambulance going to the Rockyview Hospital," he said in a clear and calm voice. Den is exceptional at triage. He would have made a great doctor. Not only because he has a photographic memory to cram fourteen years of medical school into his brain but also because he's as cool as a cucumber when other people, like me, completely melt into their pain and emotions.

"What happened?" I cried into the phone. For him to be in an ambulance meant it was serious and incredibly painful. Otherwise, he would have been calling from his truck or just waited until the morning. Like that one time when I woke up and gasped at a huge, swollen purple lip lying next to me. Or another time when the rumbling of a stranger's voice startled me into consciousness at two o'clock in the morning. One of his teammates had carried him into the house because some much younger punk (his words not mine) had pushed him into the boards so hard it pulled an already damaged lower back.

"I broke my leg," he replied unruffled. "Can you call Benji and get the car? I don't want to leave it at the arena overnight. "Oh, and don't bother coming to the hospital until the morning."

"Okay." I stuttered. "Yeah, I'll take care of it," and hung up. I sat leaning against the headboard with my arms limp on either side of my body, wondering what to do next. It was just after midnight and every part of me wanted to crawl back under the covers and go to sleep. Instead, I sluggishly scrolled through my contacts to find Benji's number and rubbing the last bit of sleep out of my eyes, I agreed to meet him in the parking lot in half an hour. I'm not sure how in all the chaos Den had the sense to leave the car keys with Benji, and that occupied my mind the entire Uber drive to the Max Bell Centre arena.

The parking lot was empty except for Den's SUV and Benji's truck. He and another guy from the team were sitting on the tailgate, sipping on a couple of beers while they waited for me to arrive.

"We didn't think it was that bad," Benji gave me a hug and handed me the keys. "He skated off the ice and we thought he'd shake it off. We called the ambulance when he passed out on the bench. I hope he's gonna be alright."

"Thanks. Me too." I pulled my jacket closed and rolled my eyes toward the clear night sky. "This is the last thing he needs right now. Well, I best get going. Thanks for waiting."

Good thing I wasn't drinking! I thought as I started the car. I waived to the boys and pulled out of the parking lot. It had been almost thirty days since my booze snooze began. The nights he played hockey were always my heaviest drinking nights because I was alone and could drink as much wine as I wanted without anyone counting the glasses. He'd only see the bottle the next day when he took the recycling out, and by then I was already fit as a fiddle again.

The next morning, I carefully opened my eyes hoping Den would be lying next to me, and it had all just been a bad dream. I

122 | Yvonne Winkler

rolled onto my back and stared at the ceiling for a few minutes. *Shit, I'll need to clear my whole schedule. I guess it's a good thing I don't have much going on right now,* I thought a bit irritably. The world I knew had been turned upside down overnight, and I needed a minute to recalibrate the direction the day was going to take.

An hour later, I walked through the doors of a busy Rockyview, thankful that they had taken him to the hospital closest to our house, saving me the stress of getting across town. With Den's computer bag in one hand and a duffle bag with fresh clothes and his toothbrush in the other, I walked up to the nurse's desk.

"I'm here to see my husband, Dennis French," I said strained. Flushed from my fast-paced walk across the parking levels in the frosty autumn temperatures, tiny pearls of sweat appeared the minute I entered the warm air of the emergency room entrance.

The nurse led me into a room where I found my husband loudly entertaining the staff. His right leg slightly elevated but no cast. For a moment I thanked God it wasn't as bad as I thought. Den, typically a quiet man, seemed in unusually good spirits given the situation. *Maybe it wasn't broken after all!* I thought, breathing a sigh of relief.

"Hey babe, did you bring my computer?" he asked smiling at me. "I have some work I need to do."

"I'm sure work can wait a minute," I said through my teeth. "Wanna fill me in first?"

As I sat on the chair next to his bed, Den began to tell me all the gory details of how his skate got stuck on the ice while his body rotated with the oncoming weight of the defense guy from the other team, resulting in a nasty spiral fracture in his right tibia and fibula. Together with the surgeon, he chose to proceed with internal fixation which I learned meant wiring a titanium rod through his knee and physically reconnecting the bones with screws.

"Geeze, Den. That sounds so painful," I said with a curled lip, slightly nauseated. Neither of us had ever broken a bone in our body and we had no idea what we were in for or for how long.

"Doc said I won't be playing again this season. But I figure I'll be back for playoffs."

"OK, don't you think that's a little overly optimistic for a pessimist like you?" I retorted.

The surgeon had laid out the road of recovery to be at least twelve months, including rehabilitation of the leg. Den thought he could do it in half that time. We knew that his body had the unique ability to heal faster than most people and often without visible scars, something we joked to be his superpower, but this? This was different. He had no previous experience with broken bones, and he wasn't twenty-five-years-old anymore. I hoped he'd see things differently once the shock wore off and be able to manage his expectations realistically, if only to spare him more disappointment and frustration.

The first couple of nights back at home were quiet. He was still on heavy-duty pain medication and slept most of the time. I went about my day as usual, checking on him regularly to bring him more water or readjust the elevation pillow. As the days turned into weeks, the agony grew. His back started to bother him from laying in one spot, he became restless, and the reality of his situation began to sink in. Not only did the accident happen on the third game of the season, but hockey was the only physical outlet he enjoyed and that helped ease the demands of his growing environmental consulting company. It didn't help that everyone he talked to, including me, cut down his dream of playing again. Most of us thought it would be best if he hung up the skates for good as the sport had left its marks on him. To Den, these well-meaning suggestions were forecasts of an impossible future coming to terms with his age, workaholism, and relinquishing control.

In the weeks that followed, our life came to a grinding halt. I was grateful that I was already working from my basement office and was able to be with him, but for Den it was like a prison sentence. Being confined to the house for two months, unable to come and go as he wanted, and relying on me to help him with day-to-day tasks like eating, showering, and driving, crushed his spirits hard and claimed his already fragile mental state. I found it more and more difficult to focus on work and manage all aspects of our family, from grocery shopping and meal planning to laundry, and taking care of a grown man who didn't want to be helped consumed all my mental and emotional capacity.

Both of our childhood trauma was rooted in lack of safety leading to a very strong fight-or-flight adaptive strategy. Outwardly that showed up by focusing on things we thought we could control like working a thousand hours, micromanaging people and projects, or continually learning so that we would never be caught without an answer, excessive eating, or dieting. We both used work to checkout from our reality. Now that he couldn't sit for any length of time at his computer, go to job sites or teach his classes, he had to loosen his tight grip on how he managed his company and trust that his staff would handle the day-to-day operations. He had to learn how to do things differently and that was not easy for him. Den's entire world was built on predictability and when that world broke, he crumbled with it.

I regulated my fear with the please and appease method, also known as being agreeable to everything, even if it caused great pain and agony that I would then drown out with booze. But since I had quit drinking, I reached for my second favorite, working on my business, or myself.

Our house became very quiet. We'd start the day in our living room, sitting across from each other in the dark. We exchanged a short "How was your sleep?" and updated one another on important appointments, then I descended into my basement office and Den would begin his daily rotation from couch to office to bed. He couldn't sit at the desk for more than twenty minutes at a time before he had to elevate his leg again to minimize the excruciating pain from the blood pooling around his ankle. At the end of the day we'd meet again in the living room, sitting across from each other in the dark. I was still licking my wounds from my perceived shortcomings as a business owner and saw this as an opportunity to demonstrate my capabilities as wife, friend, and caretaker. But everything I said was wrong, my help wasn't wanted, and my love was denied. I sat in my chair opposite the couch and opened my mouth to say how I felt, only to close it quickly unnoticed. I didn't want to agitate him, and I didn't know how to draw him out of his cave.

From the moment we began to date, Den and I kept living our lives independently. It was something we consciously decided and were proud of. Neither of us wanted to be the partner who relied on the other to make them happy. We enjoyed our time together when we had it, often slipping out of our busy professional lives to log cabins somewhere in the Rockies with no cell reception or escaping the long, cold prairie winters by going somewhere hot. For that reason, he owned two condos in Arizona, one of which, a short-term vacation rental, I managed for him and the other, a long-term rental for which we had received notice from his tenant of eight years; she was moving to Guam. We were excited at the prospect of converting this unit into a premium short-term rental and our second home during the winter months. We hired a contractor in August to help us with the big jobs like flooring and renovating the kitchen and bathroom, and we planned to go to Scottsdale in October to do the rest ourselves.

"I'll cancel our flights to Phoenix tomorrow," I said while browsing the Home Depot web page for bathtub options. Den was in his usual position, on the couch with his leg propped up on a stack of pillows. He turned down the TV volume and said prophetically,

"I should be able to go by December. I'll still be on crutches mind you, but I can paint baseboards sitting on the floor."

My eyes widened in disbelief. I didn't want to rob him of a glimmer of hope that there was an end to his prison sentence. I knew he needed something to look forward to and something different to occupy his mind. He could recite the TV rotation by heart and had finished the sudoku book I bought him when he first came home from the hospital. He was bored to tears.

"The doctor said that I should be able to put 80 percent of my weight on my leg soon. When that happens, I'm out of here."

"I know how horrible all of this is for you babe, I'm in it with you," I said empathetically. "But I can't see how we're going to get you on a flight, let alone up those stairs to the condo and then holding tools in one hand while you stabilize yourself on the crutch with the other."

I knew we both needed a change of scenery. Our relationship had always been strong because we used to have time away to miss and appreciate each other. Now that we spent every minute together, we got annoyed about every move or sound either of us made. I woke up when he did, listening for any sign that he needed help because he was still too stubborn to ask. He'd even attached a shopping bag to his crutch to transport a variety of things from room to room without my assistance. Thankfully, he never figured out how to balance a plate of food or cup of coffee on his head or else he would have done that too, just to claim his independence.

I felt so useless and unneeded. It seemed like everything was at a standstill. I needed to see a project successfully completed as much as he needed to see progress. Watching him resist support was infuriating, perhaps because it mirrored my own unwillingness to accept help. I could see how hurtful it must have been for him when I slammed the door shut on his concerns and hid out in the basement. He didn't want to see me struggle to breathe as another anxiety attack washed over me. Financial assistance was how he was able to help me but I kept denying it out of pride. I believed that being a feminist meant I had to do it all on my own. I wanted to be the hero of my story and victoriously throw my arms in the air and shout, *in your face you naysayers! Look at me now.*

***

Scottsdale, December 2018

Den eventually came around and we agreed to let our contractor finish all the work so that when we arrived in Arizona for a month in December, we would only need to focus on furnishing it. One evening, at the pub across the street from our condo, and after a wearisome day of cleaning construction dust and assembling furniture, I couldn't fight temptation anymore and ordered a glass of pinot noir with my dinner.

"Are you sure you want to do that?" Den leaned his crutches against the high table. "It's been what, three months?"

"Four, and no, I'm not. But it's the only thing that can help me unwind right now," I replied looking down at the menu. The struggles of launching my business, my ongoing inner critic yelling at me for not lifting my weight in this family resulting in constant busyness, along with the toll Den's accident had on our relationship and his mental state left me feeling insufficient as a woman, wife, and business owner, and the only thing that could comfort that wound was alcohol.

I knew that having that glass would start the counting clock all over. That's what they taught us in A.A. The moment you slip, you're back at square one. I had a problem with that but it was the lesser of all the things I faced. And so my brain justified it with a loud *fuck it, I'll quit when all this is over.* When the waiter came back with my nine ounces of wine and an IPA for Den, there was no turning back. I resisted the urge to grab the glass off his tray and gulp it down at once. Instead, I paced myself through the first sips, savoring the flavors and letting the warm tingling settle into my body.

"I think I need a couple of days away from assembling furniture," Den said as he brought his beer to his lips.

"Yeah, we've been working like we were making up for lost time," I agreed. "Want me to see if I can get us in at the Fairmont spa, maybe get a massage and hang by the pools?"

We usually enjoyed hiking the dry, red canyons and diverse Arizona landscapes this time of year. But since that was off the table as a form of recuperation, hydrotherapy and bodywork were the next best option. Den nodded and I got on the phone to make the reservations.

"Would you like another round?" the waiter asked when he brought our food.

"Yes, please," we said in unison.

After the first glass, the intervals between sips shortened considerably and by the time I was through my burger and fries, I had finished the second glass. That's the agony of alcohol. It releases powerful feel-good chemicals to the brain and acts as a natural painkiller overloading the pleasure center, triggering a craving for more of the same. Before I knew it, one glass with dinner wasn't enough anymore, I needed at least two and then three to feel comfortably numb.

The next morning, I messaged Kerry to see if she had time to talk. The guilt of having broken my sober streak was hard to bear alone in the light of day. I couldn't talk to Den about it because he had warned me about the consequences last night. Besides, this wasn't the first time and he'd grown tired of the on-off game. Den was a resolute man. Once he made up his mind about something he saw it through. I wished I could be normal like that. Only, this wasn't a test of will. This was something that eluded him because he had never smoked or had any addictions to substances, only work, and unfortunately that wasn't seen as an addiction. It was a sign of a productive contribution to the world. A couple of hours later, while waiting for another delivery, I was on a video call with Kerry.

Freedom Seeker | 129

"Angel, you need to rest! You're perpetuating the problem by distracting yourself with overwork which leaves you exhausted, wanting to reach for relief in wine." Kerry winced.

"But that's what we're doing here," I said exasperated. "We came here to get away from the stress at home."

I didn't understand what else she wanted me to do. Sit by the pool and read? When we arrived in Phoenix, we were so excited to have something to sink our teeth into we hit the ground running. To alleviate Den from strenuous trips through furniture stores on crutches, I spent hours sourcing everything we needed through an online shopping site which also delivered right to our door. The caveat to this genius idea was that we had to assemble everything piece by piece. The warm Arizona winter sun lifted our spirits and effectively distracted Den from his pain. For me, it was a new opportunity to show him that I'm a useful contributor and not just a drain on our family's income.

"Why is it so hard for me to just stay away from things that aren't good for me?" I wiped the tears from my cheek and looked out the window.

"Becoming a non-drinker is a process, angel. You're now at the stage where you have to look out for the things that will activate your old adaptive strategies." Kerry was empathetic and urgent.

"In A.A. they have a useful acronym, HALT, which stands for hungry, angry, lonely, tired," she pushed on. "When you prioritize your physical and emotional needs and take care of them, it's less likely that you'll need to resort to numbing out behaviors. But, when you neglect all four, you'll inevitably find yourself at a pub ordering a pinot noir with your burger."

I was still learning how do discern my own needs and build a self-care practice that made me a priority so that I could stay topped up all the time and not depleted. It wasn't easy to override decades of conditioning to be selfless, to suck it up, and that drinking was what strong men did to deal with problems, much like the messaging used to be that the cigarette was the feminist torch of freedom.

"OK!" I conceded. "I will try to quit again starting in the new year. But I'm not going back to A.A." I said firmly. "I don't even understand half of what that big book says, and I always feel like there's something wrong with me, that I need to be atoned for my sins."

I couldn't relate to the A.A. model at all and every time I sat in that stuffy room that dated back to the nineties with its purple wallpaper and blue carpet, I felt like a broken person. Only by admitting to being helpless and following the twelve steps, developed by a white stockbroker from Brooklyn back in early 1930s to cure his out-of-control drinking, would I be fixed. Not only could I not relate, the twelve steps didn't feel true for me. A.A., much like the world I navigated, was created by upper-middle-class white men for upper-middle-class white men. *Fuck the patriarchy.* Women don't need to be fixed; we need a community where we can be who we are, unapologetically.

In my desperate attempt to find an alternative for A.A. and its outdated framework, I resorted back to my good old friend, Audible. I found Holly Whitaker's *Quit like a Woman* and the Tempest community. I immediately resonated with Holly's direct approach to unpack the truth about alcohol, and how we got to a place where it's OK to condemn a person for not drinking enough and drinking too much all in the same evening. I wanted to be surrounded by people who wanted to connect with me and didn't need to read from a book that required a translator because it was written in the thirties.

She effectively conveyed that there was nothing wrong with me and that I fell into a very carefully designed trap that women and other marginalized members of society, fall into. Alcohol, being a highly addictive substance, floods our brain with the chemical dopamine, robbing us of our natural dopamine production, which results in us needing more. Just like other drugs, only legal and highly commercialized. Wine, it seemed, was the only thing that comforted me and that was in my control. Or so I thought. If I couldn't connect with people properly then at least let me have wine with them. Wine, just like cigarettes, was my gateway to fitting in. Without it, I'm the outsider again.

Quit like a Woman and a dozen other quit lits became my big book, inhaling the stories of women who had overcome their dependency on alcohol whenever I got the urge to drown my tired, stressed-out body in another bottle of velvety smooth oak and black currant aromas.

I tried belonging through creating success. I tried belonging through cigarettes and wine. None of them worked. I had to revisit what belonging to myself looked like.

8
EMPIRES VS COMMUNITIES

Calgary, 2019

I found a table in the corner of the Distilled Beauty Bar, a café and salon in Calgary's trendy Marda Loop area. Its modern design and eye-catching elements made it the perfect hangout for much-needed girlfriend time. The barista had just dropped off my turmeric oat milk latte when Jessica floated in through the green wooden French doors. It had become our ritual to meet here once or twice a month, ever since we both left our corporate jobs to tend to our emerging solopreneur careers.

"Can I also please have one of those?" She smiled at the barista and pointed at my mustard yellow beverage. "I could use some comfort in a cup," she added, unwinding the five-layer scarf around her neck.

"What's going on?" I inquired calmly.

"I'm so tired of people calling in sick, I can't get anything done because I have to do their work too."

Jessica had hired two women to help with the marketing and customer fulfillment components of her brand strategy business. A feat I was quite envious of as all the support I could afford were freelancers, often thousands of miles away. I was craving the bond and connection that seemed only possible with people who lived in your postal code. The disadvantage of hiring staff was that when one of them fell ill, Jessica couldn't just go and assign the work to someone else. She had spent many hours collaborating with each woman to design a deluxe client experience. Everyone brought a very specific skill set to their role. What was even more astounding to me was how she was able to sell them on the promise of a remarkable salary while raising the funds for that promise. However, that also made it difficult for her to complain about the frequent absences and made it virtually impossible to replace that member on her team.

"So, did you register for the RISE event yet?" Jessica had settled into her seat across from me, both hands wrapped around the warm mug that was supposed to lower the levels of inflammation in our bodies.

"Not yet," I sighed. "I'm struggling with how to come up with eight hundred dollars, plus whatever the accommodations will be."

"But Danielle LaPorte will be there! You love her!" she gasped. "And we'll meet some great women who could be ideal for your program, I'm sure. We could share a room, a bed even, if that helps?"

Jessica had invited me along to a few networking events where I had made some valuable connections. Going to these things always came with high levels of anxiety for me, unless I was a volunteer, exhibitor, speaker or otherwise involved with the event. The mindless chitchat and exchange of business cards that I would later find in the lobby garbage can, not to mention figuring out how to insert myself into a conversation, usually had me running for the nearest bar where I could sooth my nerves with a glass of wine. The bar would also become my main pillar to lean against and hold up my quivering legs. Networking seemed to come naturally for Jessica since, unlike me, she was energized by being around people. This was one of the attributes I found so valuable for our collaboration. I just needed an introduction and could hold my own in the right crowd. I paid close attention to how she worked the room, floating effortlessly from person to person, and by the end of the evening she'd know almost everyone.

134 | Yvonne Winkler

"OK, I'm in." I conceded, sealing the deal with a sip of my hot drink.

Three weeks later, we rolled up to a private luxury hotel in the foothills of the Rocky Mountains, and just twenty minutes outside of Calgary. We were greeted with a pink, gold, and white balloon arc display unlike anything I'd seen before. The venue was intimate, cozied up with large wooden beams that framed the floor to ceiling windows. A raised platform in the center of the room served as a stage for the lineup of speakers and was enveloped by round tables with white linen and large silver chandelier candleholders. Jessica and I looked at each other as our jaws dropped in disbelief and euphoria. We knew that anyone who paid that much attention to the smallest of details was our kind of human.

Looking over my right shoulder at the bar, I saw at least five rows of neatly stacked champagne flutes with rosé bubbles racing to the top and thought, *maybe it's not just me who relies on alcoholic comfort to break the ice.* I graciously declined the volunteer who tried to hand me a glass and followed the hostess to my assigned seat. Taking a deep breath, I scanned the room to find where Jessica was seated. *How was I going to get through this without wine or her?* I had renewed my sobriety vow at the beginning of the year and was so close to celebrating my three months milestone, I didn't want to break that dry spell here among these successful businesswomen and in front of one of my biggest mentors, Danielle LaPorte.

Danielle's work was not only foundational to provoking real change in what I was doing with my life and why, but also how I wanted to show up as a business owner and leader. Jessica had introduced me to Danielle's signature *Desire Map Planner* a year earlier. I loved her brand's clean lines, gold and pale watercolor designs, poetic words and the overall artistry Danielle brought to the otherwise stale personal development space. I consumed every post, book, and podcast she ever published but the words that forever changed me were on page seven of her best seller *The*

Desire Map: A Guide to Creating Goals with Soul. "Knowing how you actually want to feel is the most potent form of clarity that you can have. Generating those feelings is the most powerfully creative thing you can do with your life." This epiphany, that I can have everything I wanted now by generating feelings, not things, triggered a tsunami in my heart that washed away an archaic belief that only when I had the house, the money, the car, and the exotic trips would I be fulfilled. When I began to look at the world through the eyes of my heart, I could see more clearly what I wanted and what was getting in the way of it. It was hard for me to not idolize this woman. So instead of putting her on a pedestal, I put her at the helm of my imaginary board of directors that I frequently consulted for inspiration and direction.

As I sat amongst strangers in this elegantly decorated room, I wished I had a little liquid courage that turned on my easygoing, carefree, social persona. I estimated most of the women to be in their late twenties, early thirties. *Ugh, more millennials,* I thought as my mind pulled up the imaginary list of all the things they were better at than me. Younger, more creative, better at social media, inclusive, diverse, on and on. My inner critic was at full tilt, and my eyes now searched frantically for the exit sign when a soft voice came on through the speaker directly behind me.

"Good evening ladies and welcome to our first ever Women RISE event."

The organizers had done an exceptional job making me and the other introverts feel included with subtle hints about how we're all in this together, in the form of bumper sticker quotes peppered all over the walls. My favorite was the one about not hiding in the bathroom. *Why,* I wondered, *do I feel so uncomfortable and unsafe around these other women? Where did all this insecurity come from?*

"Would you like red or white with your dinner this evening?" A voice interrupted my deep thoughts.

"Huh? Oh! No wine at all, thank you. Could I please have a glass of sparkling water with no ice?" I said apologetically and immediately regretted the tone, *Duh. Why was I apologizing?* The server gave me a look of confusion, as did the woman who sat next to me.

"I'm tired and wine would for sure put me over the edge and I'd miss the keynote address tonight," I tried to explain. "Have you heard Danielle LaPorte speak before?" I inquired, looking around the table, trying to take the attention off me. "She's the main reason I wanted to attend," I proclaimed and leaned back into my chair. "I love her poetic riffs and general view on all things new age."

None of them had heard of her which I thought was bizarre, so I switched to the next question I had carefully prepared in advance to be a better networker and was genuinely curious about.

"What brings all of you to this event?"

The woman to my left was in human resources and her boss, who was sitting next to her, had paid for them to attend this event because nothing like this had been done before, at least not locally.

"Oh, that's fabulous!" I uttered hoping to conceal my envy that they were placed at the same table. I had spotted Jessica on the other side of the stage, showing off her pearly whites, clinking glasses with the other gorgeous women at the table. Small talk was laborious for me and the increased volume of chatter in the room made it difficult to hear across the table. I only got bits and pieces of what brought the other two women there. By the time we were done introducing ourselves, I felt a bit deflated as none of these women were what I had defined as my ideal customer.

Later that night, as Jessica and I sat on the bed in our pajamas reliving the night's events, I did what I always did when I felt like I didn't belong; I diverted the conversation from my discomfort to Jessica's latest pains. It helped me find my sense of purpose and worth.

"How is it going with your part-timer? Is she back to work yet?"

"No, it's so frustrating, she's on extended leave now and I promised her she'd have a place when she is better. In the meantime, I'm pulling all-nighters. I'm exhausted." Jessica shook her head and took a sip of her wine.

"I'm applying for a government grant to help me subsidize their salaries while I'm turning down clients because I can't handle more workload. Tell me how that makes any sense?"

"What will you do if you don't get it?" I asked quietly, knowing full well that she couldn't keep hustling at the rate she was going.

"I don't know. I can sleep when I'm dead though, right?" She looked at me sheepishly.

I knew how this story would end. I had heard it over and over from the business owners who visited our lodge in Nova Scotia. Many of them had built successful businesses that enabled them to finally enjoy their money with extended trips to Canada and fulfill their dream of owning a piece of land in the serene, untouched woods on the shores of the Bras d'Or Lake. But it came at the cost of their health or relationships. A few passed away from heart disease or cancer shortly after their vacation house was built. Some never even saw it before they put up the For Sale sign due to divorce. One tried to repair a strained relationship with their son, one summer vacation at a time. Heck, I had watched my own family fall apart over the stress of making ends meet. I believed that there was a better way. One where we could have it all, and where money wasn't the main driver or determining factor of success. I wanted to prevent women from burning out and going back to undesirable jobs that took them away from their passion because they thought they weren't cut out to be business owners.

Watching Jessica on that path was heart-wrenching and confusing. She was doing all that we were prescribed to do to be a successful #girlboss. But the routine of being pretty but natural, hardworking and self-caring, kind but assertive, successful and zen, on point with the latest fashion but sustainably sourced, and my all-time favorite, being an authentic leader a.k.a. pretend you have it all together, was killing us one glass of wine at a time. She didn't want my advice, and she certainly didn't want to hear that what we were doing was unsustainable, especially when it came from someone who was doing all she could to keep up. I felt this enormous urgency to hurry up and show her and others that there was another way, and it had nothing to do with money or fame.

"I signed up for B-School." I beamed. "Danielle LaPorte did this affiliate thing with Marie Forleo where she offered the Desire Map Facilitator license for free. I figured that maybe if I aligned myself with an established brand like hers, it would give me more credibility and access to a community of women who think like us for my group coaching program."

"That's a great idea! I heard lots of good things about B-School too. Actually, one of the pop-up vendors here this weekend went through that program and became a mentor for Marie's community. Maybe that will give you even more lift." Jessica seemed genuinely happy for me but I could tell she wasn't interested in talking about my business. She only had the capacity to deal with her challenges and getting that grant to help her advance to the next phase. Having Jessica in my life alleviated some of the loneliness I felt as a solopreneur. Every decision that needed to be made had to be made by me. I was the creator, writer, marketing, sales, recruiter, and coach all-in-one. That wasn't her reality. Jessica had tons of people always buzzing around her hive, making me feel like I was just another worker bee in service of the queen. I had hired Jessica's company to help me with my brand strategy and website design. For me it wasn't a question if she was the best for the job, she was my friend and fellow start-up business owner so the natural choice. "Together we thrive." Isn't that what a group of top businesswomen said at the panel session we attended tonight?

Freedom Seeker | 139

Hiring Jessica was helpful in many ways and a huge relief. I needed reassurance and someone who could see what I couldn't anymore after editing my profile page for the hundredth time. I wanted what she had with her girls. Business meetings in trendy cafés where ideas and solutions came alive. But it wasn't just marketing and content. I was on a steep learning curve. Instagram was just taking off in a big way and the marketing options that became available to us through various social media channels were enormously complex.

Jessica was everything I wanted to be young, popular, and creative. She knew people and people knew her, and when she spoke everyone seemed to listen. As students of the divine feminine rising, we shared a passion for a heart-centered world. We had similar taste in visual design, and she understood why it was important to me that my business cards were printed on soft-touch paper stock and why taking time on deciding the right font mattered. We were attuned in almost every aspect of how we wanted to impact the world. Torn by the dissonance of what I knew about the patriarchy, the path to success, and what she had, I realized in that moment that I couldn't be like her and make the impact I wanted to make in the world.

After the final session of the retreat, we all flowed into the reception room where Jessica and I had been welcomed just forty-eight hours earlier. The bar was now lined up with wine and a mouth-watering assortment of cured meats, various cheeses, olives, nuts, fruit, baguettes and crackers. The DJ, a fierce, charismatic woman with the same undercut as P!nk, tended the turntable and everyone, including Danielle LaPorte, were up on their feet dancing, throwing their arms in the air and singing along to the music. I had met some amazing women. There was Kristi, the owner and founder of Encircled, an ethical and sustainable clothing company out of Toronto, and Robin Joy, a soulful, radiant yoga teacher, passionate about supporting women to have a deep relationship with their bodies and pleasure. Standing in line for

Danielle to sign my copy of her book, I met Stefany who had recently left her nine-to-five at a bank to follow her true calling as a holistic nutritional consultant. I never expected to meet this many incredible leaders, so genuine, accessible, and generous. As I looked over the crowd of bobbing heads, it dawned on me that networking wasn't about finding the next client. Networking was about making connections with people who share a common goal. Being surrounded by women who weren't here to compete with each other, and observing Danielle LaPorte's leadership in action, showed me what was possible when we serve from the heart and let down our guard. My social anxiety dissipated with every speaker and transformed into a sense of collective purpose. I didn't need a drink to get through anything and I felt like I belonged, until I didn't.

The RISE event gave me a renewed energy to move confidently into the emerging spring season. The last of the snow had melted away in Calgary and the lilac bush in our backyard was beginning to bud. Soon all the trees would be bursting with new, lime green leaves, the robins would return from their winter migrations, and the swallows and wrens would get busy forging their nests in the birdhouses we hung along our fence. I loved listening to their mating songs from my kitchen table in the morning where I had set up my temporary office, giving this "work from anywhere" slogan a try. It seemed every creature was busy getting ready for new life.

I too was optimistic that this year would be the year my business was going to turn its first profit so that I could hire my first employee and begin to pay back Den. That March was the first anniversary of the group program I had developed, and I planned to relaunch it with a new format and update it with all the new insights I had learned. Still flying on the tailwind of the event and inspired by the women I met, I figured the key to my success was in coming out from behind the keyboard and to get visible. I lined up a few new interviews for the Freedom Seeker Show, and, just as

Marie Forleo and Danielle LaPorte ingeniously modelled with their collaboration, I wanted to connect with one or two business owners who corresponded with my values and served the same audience, but with a different solution than I offered.

***

I pulled into a full parking lot at The Winston Golf Club and spotted Anne's striking curly, silver hair right away. We had become acquainted over the last year at several business networking events. She seemed to know and hit them all, like Marla Singer in the movie *Fight Club*. I was inspired by her taking advantage of this resource to promote her travel business because I still hadn't found one where I felt like I belonged. Thinking there was something wrong with me, I read books on the subject, hunting for clues on how to correct this flaw. In her book *Networking for People Who Hate Networking: A Field Guide for Introverts, the Overwhelmed, and the Underconnected*, Devora Zack taught me that I didn't have to be extroverted to be a good networker, so I agreed to join Anne at one last meeting. She had been a member of this group for a year, give or take, and raved about the quality of topics discussed, the people she met, and the graceful way she met them.

"Hello, Yvonne," a tall woman greeted me with a subtle European accent. "I'm Yvonne Basten, Managing Director for eWomenNetwork, Calgary chapter."

I'd only met a handful of women who shared my first name, and it always made me feel excited, like sharing a unique name comes with an immediate and intimate knowing of that person.

"You're at table five, go ahead and get settled, we're about to start." She beamed

"Wow! We get to sit at the same table, that's great!" I hooked my arm into Anne's and meandered through the naturally

142 | Yvonne Winkler

lit room with a fantastic view over the golf course, stopping only briefly to say hello to someone Anne knew. *Ugh, I really need to make a point of arriving earlier to get a seat so my back isn't to the front of the room,* I thought unnerved as we sat down in the only two chairs left at the round table. Yvonne opened the summit with a short introduction to eWomenNetwork for those of us who were new, followed by the organization's nine values, and then we had few minutes to go around our table to meet the other participants. This was a guided process and being limited to meeting five new people at a time was definitively easier than being unleashed into a room full of strangers. Anne was right, this group had a different vibe. After we introduced ourselves, I grabbed a cup of coffee and settled back into my chair, legs crossed again but unclenching my arms to listen to the keynote speaker, Sandra Yancey.

I'd never heard of her and was instantly curious when I learned that, despite being the CEO of this multi-million-dollar organization, she took the time to visit small town Calgary in the spring, a season that's generally unpredictably cold and grey.

I estimated Sandra to be in her late fifties, early sixties. She knew how to command a room very organically and authentically. Big hair and short in stature, she was the classic image of a woman from Dallas, Texas. Being raised in an immigrant family with language barriers, and financial and racial hardships, I could relate easily to both her drive and determination in what she had built. When she spoke about how she started eWomenNetwork twenty years ago, when few recognized the need to support women entrepreneurs or the importance the digital world would play in everyday business, I placed both feet on the ground and leaned in. I found a renegade who, like me, questioned the systems and structures of the world without any certainty whether her risks would turn out. The women in the room were also stirring in resonance, some drying the corners of their eyes when Sandra talked about the lessons she had learned from her mother. She delivered a powerful message about how we were all leaders and can absolutely make our mark in the world, but we can't do it alone. "Lift as we climb" is one of eWomenNetworks' nine values.

When Sandra finished speaking everyone raised to their feet and gave her a standing ovation. Being in that room with a hundred women ranging from start-up to established business owners, all between their early thirties and late sixties, I felt like I had found a community I wanted to be a part of. I needed a network with ongoing collaborations, and this organization was well established and shared my values. Without further thought, I grabbed the sign up form from the center of the table and joined one of the largest and widely recognized premier business networking organizations.

The line for the lunch buffet had finally died down a bit and most of the women were seated back at their tables eating and having lively discussions. With a deep breath I tried to digest the last few hours of information and get myself ready for the mingling and socializing portion of the event, when a shrill woman joined the queue behind me.

"Hi, I'm Angela." I turned around to meet a tall brunette who stuck out her hand. "What is it you do?" I looked over the buffet options, hoping she wouldn't see my disappointment. *What ever happened to small talk? Always straight to elevator pitch.* I thought grimly.

"I help women create better lives for themselves," I said with a forced smile and grabbed a plate.

"Oh, so you're a life coach?"

"Yes, aspects of life and business coaching are involved. I find it's hard to separate one from the other," I said. I didn't like labels but I understood the need for them for people to compartmentalize.

"So, you work only with women? That's really interesting. What made you decide that?" She wasn't letting up.

The woman in front of me now turned, scanned me from head to toe, and with a snooty tone chimed in, uninvited.

"I think that's the problem with feminism, it segregates us women more by putting a special needs sign around our neck."

Pretending to not be perturbed by this grey-haired lady in a red blazer and matching pencil skirt, I rolled my eyes, stabbing around in the lettuce bowl. I looked up to find everyone had stopped, waiting for my reply. I took a deep breath and tipped my head in curiosity.

"Why do you feel that addressing the struggles unique to women is a problem?"

"I've been a business coach for twenty-five years and I can tell you that it's the segregation and special treatment of women that perpetuates the problem. If we want to be treated equally, we can't be making special requests." She turned back to fill her plate.

I felt a warm tingling crawling up my neck as I thought of all the things I wanted to shout at her but seeing the women around me, I concluded it was unnecessary. They seemed to get why we were at an all-women networking event. I moved the line forward and replied, "Well, that's what we have been conditioned to believe in order to uphold the patriarchy." I grabbed a piece of steamed salmon and walked back to my table, flustered but satisfied on how I handled this situation.

This wasn't the first time I encountered resistance to my vocation to support women exclusively. I spent my entire career fixing myself to become more like a man, ditching my feminine qualities so that I appeared stronger, less emotional, to gain acceptance amongst my peers. Once I had built my empire, I could use that money and influence to affect the change I wanted to see in the world but first I must make it there. That was the model I had followed and it only left me physically, mentally, and emotionally exhausted.

Like the day my student councillor told me I need not bother to apply for university, this woman only affirmed for me what I already knew. We needed to heal our relationship with each other before we could create equality and a better world for the marginalized members of society.

"I liked what you said back there," Angela had found me and sat down next me with her lunch. "We need to stick together. So how do you help women make a better life for themselves?"

"Well, I start with creating a safe space where women can explore what creating a sustainable future that benefits everyone could look like." I shoved a fork of greens in my mouth to slow myself down. Angela nodded which I took as a sign for me to continue. "I've spoken with hundreds of women throughout my eighteen years of working in sales, customer service, and business development and the story was always some variation of finding balance between the traditional roles assigned to us by society and the work we felt called to do." I paused to take another bite. "After two years of living through my own version of launching a coaching practice, I recognized that one ancient truth was undeniable: we thrive when we work together. We need to unlearn the competitive nature of business and place emphasis on collaboration because there's nothing fulfilling in doing it all or by myself."

"Yes, I can certainly attest to that!" Angela raised her right hand.

"We learned to be pleasers and appeasers, and most of the women I know operate from a place of scarcity (a.k.a. "I am not enough") because we have been ignored for centuries." I was fired up, fueled by the scene at the buffet table. "We don't feel that we belong or that we're enough because patriarchal values and male-centric social customs have demonized feminine values for over five thousand years. So, we learned to suppress our essential qualities to 'fit in', often playing up masculine values." I let the waitress clear my plate. I had so much more to say but I saw Yvonne make her way back to the podium.

"You seem very passionate about your work." Angela observed.

"I grew up watching my mother, grandmothers, aunts, and other women in my life prioritize their husbands, families, or careers, working really hard because nothing was ever enough, self-sacrificing and ignoring their own needs. Insisting on matching what they were feeling inside about value with their outer circumstances; burning out, bitter, resentful, and depressed. I've always had a strong sense for inequality and have been subjugated and undervalued for far too long, that is why I decided to work with women."

Driving home after the event, my head was buzzing with thoughts, excitement, and possibilities. Angela was right. I was very passionate about supporting women and shattering the archetype that women are catty, reinforcing distrust and competitiveness. The event left me feeling more assured in my purpose and where I could make the biggest impact. This seemed like a great business community with members who were open to a different paradigm, except maybe the one lady from the buffet line. *I wonder what it takes to lead a chapter,* I thought, turning onto my street. Just like I had done in my former years, I wanted a closer look behind the curtain and learn from the woman who created this empire.

The next day, while out on my newly established routine morning walk through the paved pathways around our neighbourhood, I phoned Kerry, full of excitement.

"Being hitched to an established organization like eWomenNetwork would give me credibility, training, and an established community of women who want to break free and create real change in the world," I said in one breath.

"It sounds like a natural evolution of your work." Kerry echoed.

"And it would help me grow my sphere of influence exponentially." I added.

Freedom Seeker | 147

Taking some time away from desperately trying to create, sell, and deliver a program, I realized that I couldn't make the impact I wanted without the help from others. I needed the support of other business owners to grow a platform where I could share my experiences. I had created a ninety-day program and a community based on nearly identical values as eWomenNetwork's, but nobody knew who I was. This could be a great way for me to demonstrate my leadership ability and set myself up as a credible coach in the community, fostering local over global resourcing, which I believed to be one of the shifts necessary to stop the depletion and, frankly, unsustainable economic system we all participated in.

Having been raised under communism, solidarity was ingrained in us early on. We didn't have much which made us very resourceful. My mom, a talented seamstress, sat for hours into the night finishing up the hem on a friend's dress or making a colleague's daughter's wedding gown. In exchange, she got first dibs at the local shoe store (where the daughter worked) for a special and highly sought-after pair of red patent leather shoes for my first day of school. When my parents built our house, neighbors lent a hand with whatever skill set they had. Someone was an electrician, another a painter. One was a bricklayer and others helped with babysitting the kids and supplying the troops with food. Thank you, Oma Emmi.

Naturally, I always looked to support my friends and local business owners whenever I could. I also believed that this was a much healthier and more sustainable way to keep economies going. By sourcing our produce locally we supported the farmer, his family, and the people he employed. They could now buy what they needed at the local butcher and so on. Since prices were controlled by supply and demand, everyone benefited when everyone participated. In East Germany there was no choice about this. We didn't have free trade or import and export agreements. We were a self-sustaining country by design.

Being part of a business network that valued and encouraged business owners to support each other seemed like a smart first step to closing the gap, albeit rooted in the United States. I hadn't found anything as established locally. With Kerry's blessing, I reached out to Yvonne Basten, who was ecstatic to hear I was interested in becoming a managing director. My timing was divine. She had been thinking about turning over the role to someone else. She had managed the Calgary chapter for ten years and wanted to focus on her own business now. Equipped with a new role, and a renewed vow to stop standing in my own way, I was ready to facilitate masterminds and workshops and show women the power of gathering with other women and looking into our hearts for answers. I opened a new Google document and typed Business Strategy 2020. I could feel a breakthrough ahead.

***

Den and I decided to skip our usual February beach vacation at the beginning of 2020 because we had just gotten back to Calgary from staying in our Arizona condo for the month of December, and I was eager to dig into my new role as managing director and a well-formulated, and hopefully profitable, business strategy.

"Aren't you glad we didn't go on a cruise this year?" Den said as we were preparing dinner on very cold February evening.

"Ugh, I don't know. Plus thirty beats minus twenty any day for me. Why? What happened?"

I didn't pay a lot of attention to the news. Never have. On the other hand, Den watched both the morning and evening news every day and then filled me in on the highlights, saving me the hassle.

"A virus broke out on a cruise ship and it's so contagious that they've locked down thousands of people on the ship. They called it coronavirus and apparently it started in China. You gotta see this!" He turned up the TV and took over with cutting the onion.

I wiped my hands on a dishtowel and captivated by the images, sat down in one of the armchairs by the living room window to watch the scene unfold in front of me.

"The Diamond Princess cruise liner places thirty-seven hundred passengers and crew under mandatory quarantine after ten people tested positive for the coronavirus." The news anchor announced.

There had been so many deadly catastrophes in the last couple of years, from raging fires burning the Amazon, cyclones, heat waves, and floods. In 2018, a measles outbreak hit record highs in Europe, Ebola spread through the Congo, and the norovirus caused havoc during the 2018 Winter Olympic Games in PyeonChang. Seeing horrible things happening on our planet was sadly not that shocking anymore.

I watched bewildered as a young American man, who was living in Wuhan, China, was interviewed about his experience of the city's lockdown. He described how he had to pass checkpoints and was only allowed out to pick up food. He was extremely concerned about getting caught filming the barricades and army presence. No cameras were allowed. All exits closed. This story tugged on a nerve.

A few weeks later, Canada, like the rest of the world, came to a complete standstill and overnight Den and I were once again confined to our house in Calgary.

"If you get a call from the police one day that I've been arrested, please come and bail me out." I put the heavy bag of groceries on the kitchen counter and blew a strand of hair out of my face.

"Why do you keep doing this to yourself?" Den got up from the couch to help me unpack.

"Well, we need to eat. Unless you want to go to the store next time, there's really no other choice."

When the mask mandate was introduced, I was surprised to find out that I relied on reading lips more than I realized. It frustrated me no end to not be able to hear or understand when people talked to me. When the cashier at my local grocery store yelled at me through the plexiglass in a thick middle eastern accent for putting my groceries on the belt before she'd had a chance to clean it, anger of epic proportions bubbled up, and it took everything I had in me to walk out of the store without punching something.

"I can't believe the anger that is coming up for me. I barely recognize myself. I mean I know it's not her fault and she didn't make the rules, and yet I find it nearly impossible to control my rage," I said still vibrating.

I didn't know if it was the lack of human connection, the capacity line up to get into the store, the two feet distance we had to keep from each other, or the stupid arrows telling me the direction I had to follow down near deserted aisles with several empty shelves. I was triggered in monstrous proportions. A sudden flash from the past entered my mind. I remembered a visit from my mom's friend Margit, the kind aunt from the west who occasionally sent us Toffifee and Kellogg's Corn Flakes when I was nine. She had taken the train from Nürnberg to Leipzig to visit us in East Germany and described in horrific detail the embarrassing moment East German border guards made her empty the entire contents of her suitcase in front of the other passengers on the train, looking for and confiscating anything that could give us a taste of freedom. She brought us things we couldn't buy in East Germany. The shelves in our little town's Konsum store, dusty and bare. Mom stood in line for an hour to get three oranges for our Christmas stockings. I didn't like going to that store with her. I was grossed out by the stuffy smell and the bologna behind the butcher's glass already had green edges

"And to make matters worse, every time the public service announcement with the Chief Public Health Officer, what's her name?"

"Theresa Tam" Den replied, putting the milk in the fridge.

"That's it. Tam. Whenever they play that god awful message, I feel like I'm in an episode of *The Handmaid's Tale* or something." I sunk into the chair and looked at Den as tears welled up in my eyes. "Am I losing it? I feel like I'm losing it."

I felt like my husband and my parents were the only people I could express myself to without being judged. Everyone else just immediately labeled anyone who didn't conform or questioned the COVID regulations as a bad human, anti-masker, conspiracy theorist, or just generally uncaring.

"Den, I feel like I'm living a horrible déjà vu. This is like East Germany where we had to be careful about what we said out of fear that a neighbour would report us. Then the secret police took those people who disagreed or wanted to leave the country away, declared them insane and sometimes they simply disappeared!" My terror mounted with every word.

"Once the weather warms up we'll at least be able to get outside and meet up with our friends and family again," he said with a soothing tone. "Unfortunately, this won't be over for a while, so you need to figure out how to deal with it."

I knew he was right. I had to sort through the emotions. I couldn't let fear run my life again, it had taken almost thirty years to rewire my nervous system whenever I saw a cop car or passed through customs. But it felt so real. Everything we were experiencing now was a version of a life my parents and I had given up everything for to escape. I went to the wine rack, pulled out a bottle of pinot noir and poured myself a glass. This was how I knew I could deal with it, bringing my almost three months dry spell to an end. Once again, I self-medicated my pain. It was a good thing it had been six years since my last cigarette or I may have lit up with my drink. *Tomorrow I'll talk to Kerry. She will know what to do,* I reassured myself.

***

Kerry was able to squeeze me in a couple of days later at her house. She was considered a mental health worker and therefore thankfully exempt from the in-person rules. She opened the door and stepped back. No hugs, just a gentle request to sanitize my hands and continue to keep my mask on. It was clinical and weird but way better than another Zoom meeting.

We skipped the usual "How are you?" and I went straight to verbally vomiting all my anger onto her red, distressed oriental rug.

"What would you like to be different, angel?" Kerry poured me a cup of tea and sat back in her Bergère chair. She had dissolved her downtown office when everything was shut down and set up the purple sanctuary in her living room. Her temple dog Toby, a gentle giant Sheepadoodle, was laying at my feet, a show of his affection at my distraught state.

"I wish I could trust the information that we're given," I said after a few seconds. "There is so much information out there and I'm afraid the government has lost my confidence and trust by the slithery way they handled everything so far. I don't know what to believe anymore and the distrust is making me feel unsafe and that makes me angry."

"Yes, angel. You've been activated by your childhood trauma. All the effects of the COVID experience on the nervous system felt like what you, your family, and everyone in East Germany felt," Kerry said understandingly.

"But it's not the same experience now. The nervous system stores the experience of fear, and because fear has been activated by the circumstances and careless languaging your brain wants to act accordingly. It doesn't know the difference. Because it feels the same in the nervous system we must act accordingly. Fight, flight, or freeze."

I began to see what was happening. "So, I chose freeze when I uncorked that bottle of wine?"

"Yes, angel. What we need now is to come together and take care of each other. But coming together is deemed unsafe, only amplifying the collective fear."

Kerry, normally calm and self-composed, looked tired. I wondered what trauma was activated within her when she pressed on.

"What's happening with eWomenNetwork? Are you able to connect with other women there?"

"We pivoted all our in-person meetings to Zoom. Sandra felt it was important, perhaps now more than ever, that we continue to have a place to connect, be seen and heard. And small business owners are hit the hardest with all the shutdowns."

"And how is that for you?"

"I think it's ironic that I became managing director at a time when we have to go online. If anyone would know how to run webinars, it's me. I've been doing them for years," I laughed. "But I joined to get away from behind my computer and make some new connections, and that's not as easy as it is in-person," I added.

"And your training? You won't be able to go to Dallas now, right?"

"Yeah, the airline gave me a flight credit for the future. But not meeting the other managing directors in person is a huge disappointment for me," I admitted. "I really feel like I'm on my own again."

Alleviated by our conversation, I drove home and locked myself into my office to organize a six-week online Desire Map retreat, free of charge, for anyone like me, who needed to connect, feel their feels, and reprioritize their future. We needed something to look forward to and I wanted to give back what Kerry had offered me. A safe place where I could look at what was happening. While isolation and social distancing were the primary message on all media outlets, I agreed, we needed to come together, and I knew how to make that happen safely via Zoom. Kerry had once said that the way I empower myself was by healing my trauma, and the way to heal my trauma was in community.

With a renewed purpose, hope, optimism, and resources, I smiled into the camera and facilitated eWomenNetwork meetings for the dwindling group that continued to show up. This was where my lived experiences and developmental work met. I was made for this, but I couldn't lead or make the impact I wanted if I kept getting muddled up in the side effects of numbing out and was unable to be present to hear where my inner wisdom was guiding me. I looked up from my desk at the unmarked glass dry-erase board above my head, grabbed a black marker and wrote in big bold letters: "It is time to sober up. The world needs you!" No matter where I sat in the room, these words anchored me back into my mission.

The weather began to warm up outside, not exactly a gradual transition in a city at the foothills of the Rockies. Most of the snow was gone by Easter, and we welcomed a new addition to our little family. Sophia Sady was a mighty, chocolate parti Miniature Schnauzer. The previous fall, during Annie Grace's thirty-day *The Alcohol Experiment*, I had decided that a small dog would greatly benefit my journey to freedom from alcohol and registered my name with a breeder for the spring litter. Den was less than excited about it. He thought a dog would restrict our lifestyle. I disagreed, firmly. I knew in my heart that she would bring us joy and force us to get outside every day, twice a day.

Sophia is a classic Greek name meaning wisdom, and she was going to be my reminder to take care of myself and to do what it takes to get and stay sober. I didn't quit drinking the moment we got her but she effectively took our attention away from the constant COVID numbers regurgitation on television and even got Den to spend more time in the backyard playing tug, catching a ball, and giving belly rubs. She was the missing piece for me to feel whole and complete as a family. Her little paws on my thighs and her puffy tail wagging, reminding me to look up from my screen and get her a treat. She was also a natural conversation starter. We met so many of our neighbors that summer, relieving some of our social isolation, and her unconditional love was what my soul had been missing since our golden retriever passed away twenty years ago.

As summer turned to winter, like most of our family and friends, we adapted to the new normal. A life consisting of infinite screen time from workouts to conference calls, streaming shows and shopping. Even the puppy training happened over Zoom. The winter was the most challenging time for us. With temperatures plummeting to minus twenty, getting outside for a walk was limited. We chased Sophie up and down the stairs to get our steps in and to activate her bowls so that she would complete her business in the eight minutes we had before frostbite could damage her exposed paws.

In January 2021, the whole world sat on the edge of their seats, waiting to hear what changes would come upon us now with a vaccine on the way. I had a very strong suspicion that after the temporary easement of restrictions over the holidays, which usually triggered the numbers to spike again, they were going to lock us down once more.

"The feds announced a mandatory fourteen-day quarantine for anyone coming back into Canada, with a minimum three-day stay in a government authorized hotel at the traveler's expense up to two thousand dollars." Den announced as we drove down Sarcee Trail on our way to the dog park. Big fat tears welled up and my body began to shake almost violently. I had flashes of armed men in uniform standing at border crossings, holding their rifles, ready to shoot anyone who tried to make a break for it. I recalled my mom's older brother, Uncle Harry, who had fled his home in East Germany just days before East German soldiers laid down a fifty-kilometer barbed wire barrier overnight to close off access between East Berlin and West Berlin for forty years.

"We have to get out of here!" I was frantic. "I will not be on the wrong side of the wall again."

My reaction surprised Den. I surprised myself by my visceral response to a fear I knew wasn't logical or real anymore.

"I can see how difficult all of this has been for you, babe." His eyes didn't leave the road but his hand reached over to find mine. "We'll be OK, love. We'll find a way."

***

When I met Den, he made no bones about his dislike for Calgary. He had wanted to move away from Canada since he was a child. With some family roots in the US, he dreamt of faraway places, a real-life Walter Mitty who was tied down by his morals and familial duty. We often talked about my experience of moving to new cities and countries, comparing notes. I had hoped he would see that moving to a new country can be challenging, especially when it involved a completely different climate and language. But, after twelve months of being shut in and shut up, I was onboard with the idea of finding a new home, a simpler life with less noise and definitely warmer weather, anywhere!

One Sunday morning, Den turned off *Canada Over the Edge*, a TV nature series we had started to watch in place of the news and began our old familiar game of ping-pong the reasons for moving. We agreed that if we were stuck in Canada for another year, without the ability to travel to our usual escapes, we wanted to at least make a home where we would feel for most of the year the way we did on vacation, and quench our wanderlust only when we wanted to and when it was possible again. The occasional motorcycle trip to Arizona, Oregon, or California and once a year maybe to one of the Caribbean islands we hadn't explored yet. Since the United States was still off-limits, we decided to explore

British Columbia more seriously, from the dry interior to the coastal west. We would live in a few places for a month or two to get a feel for the community. *Good!* I thought. I can do that. No commitments until we know we've found the magic place.

The methodical planners that we were, we asked our relator, Sarah, to get a listing ready, just in case we were ready to sell and optimize our return on investment in a hot real estate market. I began to search out a few Airbnb options. It was only the beginning of 2021 but due to the existing travel limitations, vacation rentals within Canada were in high demand.

We finally booked our first two months in the small town of Nelson, British Columbia, where we lived in a two-bedroom vacation rental overlooking the west arm of the Kootenay Lakes. It was April again and spring had sprung. While the area was in a snowbelt, with a great ski hill, spring in Nelson was an actual season. Yellow and purple crocuses emerged rapidly from the last areas of snow, and the smell of burning brush hung in the air as locals cleared their properties of the winter debris left behind by the strong winds and heavy snow.

I continued to facilitate two eWomenNetwork meetings every month via Zoom, an endeavor that left me increasingly drained. I had only met a few of the members in person at the summit event where I decided to become a member myself. Hosting a business networking meeting solely online came with its own unique challenges. At first there was a big resistance to meeting on Zoom which led to fifty percent of the members pausing their membership. Many business owners were struggling to make ends meet and had to cut expenditures. Some closed their doors completely. With more membership cancellations than new enrollments coming in, the group thinned out substantially and we lost dynamism. Parched for motivation, inspiration, and connection, I attended the managing directors meeting every Monday and Friday, but even there the toll of the pandemic was apparent as one by one women resigned. Most of all though, I was disappointed

158 | Yvonne Winkler

about the direction the networking organization was going. Virtual meetings opened the doors to other chapters across North America, which challenged one of my fundamental values, local before global. We basically networked ourselves out of business and encouraged competition rather than supporting each other in our community.

Sharing the kitchen table as our office in our temporary home in Nelson gave Den new insights into my workday and the number of hours I spent treading water.

"I'm concerned, babe," he said one Friday afternoon as we sat in the multicolored Klondike chairs on the metal dock, soaking up the spring rays and watching Sophie explore the sandy beach. I lifted my gaze and tilted my head inquisitively.

"You're glued to that computer 24-7," he continued. "Hopping from meeting to meeting, and none of it is lighting you up."

I wanted to raise my hand in objection but stopped. Yes, I was miserable. I kept busy to distract myself from what was happening around us, and the only impact I was making was on my health and wellbeing, and not in a good way.

"What do you suggest I do? Walk away from all of it?" I said half-jokingly.

"You don't have to shut everything down but maybe park your business for a while and catch your breath."

Park my business? That was not a conceivable option. It sounded like defeat, and I was not ready to walk away from all the sweat and tears I had poured into it.

"What are you going to do when eWomenNetwork meetings go back to in person?" he asked.

I had been thinking about this for a while myself. There was enough buzz in the air that restrictions could lift by summer, in which case I wasn't going to be able to manage my chapters remotely anymore. Den's phone rang and jarred us both back to the here and now. It was Sarah, our relator, who called to tell us that she'd been keeping a close eye on market trends and if we were still interested in selling, now was a good time. The weekend was shaping up to be full of some big and difficult decisions.

I did the math. Over the past twenty years I had moved on average once per year, so I wasn't a stranger to moving. But this was different. This was the first house we owned together. It was where Den popped the question and where my dad helped me build a beautiful garden. Our backyard used to get a lot of oohs and aahs from people walking by on the path it bordered. Den had just finished a wine room to be proud of. We had built ourselves a beautiful home.

As I wrote in my journal I reflected upon the events of the past five years. Den had sold his environmental company. I had a coaching business that, up until this point, served mostly as an expensive education on what not to do. We now had a dog, and we realized, particularly through this last year, that Calgary wasn't the climate or place we wanted to spend most of our time. *Was now the time to make the change we had been talking about for years?* I wondered. I knew we would find another beautiful place but at what cost? Then there was the part of leaving our family behind. No matter where we'd end up, it was going to be more than the twenty minutes or three-hour drive away. Perhaps the biggest sticking point was something Kerry had said when I told her about our plans to sell the house and travel.

"You're reacting to fear and doing exactly what your family did in East Germany, getting away from the danger." I heard her soft voice in my ear.

Were we running away from fear or moving towards our desires? This felt more like what my *back to my roots and beyond* trip initiated. We wanted to change our life and the style in which we were living, and everything pointed in the direction that this was the time for us to do something about it. I closed my journal and joined Den in the living room of our little vacation rental above the garage and said,

"Let's do it. All I ask is that I have at least three months to clear out. Unless someone wants to buy the house as is, furniture and all."

Den looked at me with a big smile. "Really? We're doing it?"

***

One week later, the call came. "Well, it's official guys! You just sold your house." Sarah cheerfully yelled into the phone. My stomach dropped a little, my throat went dry. Gulp. A familiar feeling of the "holy shit" moment arose in me as I looked over at Den who couldn't stop grinning.

"Oh and…" Sarah paused for effect, "they want all the furniture, utensils, dishes, everything!"

I took it as a sign from the universe that we were on the right path. Ease and flow.

9
GOING HOME

East Germany, 1981

"I'm going home," I said, my small hands tightly gripping the straps of the baby blue, terrycloth backpack hanging loosely over my shoulders.

The woman who appeared in front of me looked like a guard with her uniform and mop stick. It was as if she were guarding the double doors at the end of the black and grey speckled granite steps leading out of the children's ward of the Peniger Hospital. She wiped her hands on the candy-colored Dederon overalls and with a grim chuckle said,

"Not today you're not and certainly not on your own."

I shuffled my feet along the painted cement floor as fast as I could, with her big hand tightly around my small wrist as she hustled me down the empty corridor and back into the crowded room where the other kids were still sound asleep for an afternoon nap, the perfect opportunity for a quiet escape out of this cold and ugly place. She lifted me into the crib and pulled up the metal bars to ensure I couldn't climb out of it again. Laying there in my small cage, I stared out the tall, narrow window and watched the branches of the giant oak tree sway with the wind.

It had been three days since Mom and Dad had left me there after another long night of stomach cramps and vomiting. There were no obvious reasons for these symptoms, so the doctor on duty admitted me to undergo tests.

When my parents arrived that afternoon, Mom had heard about my attempt and inquired in her gentle motherly way why I wanted to break out. I told her that the other kids were mean to me and that I didn't like porridge.

Going home to me then meant being true to myself and being accepted for who I was and not trying to fit in with the mean kids. If only that freedom seeking toddler had remained.

***

Calgary, May 2021

We arrived back from our two months stay in Nelson, BC, at the beginning of May.

"I need your help!" I pleaded with my friend Lisa, a professional organizer. "We just sold our house, and we've got two weeks to pack up our life and put it into storage."

"I'll be right over!" Lisa replied.

Twenty minutes later, I sat on the floor in our recently finished basement going through my treasure box of memorabilia of moving house. I pulled out four thick yearbooks and brushed my hand over the silver St. Francis Xavier University logo embossed into the navy hardcover. *Do I really want to move these again?* I pondered. I wasn't in any of the hundreds of photographs featured, except for the year I graduated. Much like in high school, I was never the social butterfly, preferring a quiet night in over being the life of the party.

"She's downstairs," I heard Den say followed by footsteps coming toward my hideout.

"Hey, Lisa. I'm so glad you're here," I wiped my nose with the back of my hand and got up to give her a hug.

"Want some wine?" I didn't wait for her to answer and pulled out another glass and handed it to her. "Thank you for coming over right away."

I couldn't look at her. I knew if I did, we would both start bawling. Our friendship had just begun to grow on our trip to Bali together, and now I was leaving to who knows where? She pulled out a blank calendar page she had printed prior to coming over, and looking around the pile of boxes I had already pulled out from under the stairs she said, "So let's work backwards, when do you need to be out by?"

Grateful that she didn't ask any more questions than what she needed to know to help me get organized with the task at hand, we strategically moved through each room and allotted the number of hours it would take.

"I've moved so many times in my life but none ever had me this paralyzed," I said to her when we arrived upstairs at the kitchen table. "I feel like I can't make even the most basic of decisions right now. So thank you, this was helpful."

After Den and I stored our personal belongings into a ten-by-ten storage unit and loaded up our oversized SUV with essential items we needed for the next unforeseen number of months traveling through Western Canada, there was only one thing I still needed to do. I needed to come clean. Uprooting our life in this way was like going into the attic to find that final box and wiping off the thick dust from thirty odd years. I vaguely remembered what I had carefully stored away, to never really forget but put out of sight. Now, with ten years of personal development work under my belt, I felt prepared to face the one thing that was robbing me of the freedom I so desperately wanted. My addictions.

This was a fresh start, and I was tired of doing the same things but expecting different outcomes, yo-yoing my way through life. I was tired of constantly being in recovery or working on relapses. I just wanted to be free from it all. Not having to think about whether I should or shouldn't have a glass of wine. Hiding the empty bottles from my husband and feeling the shame of being obnoxious around our friends. Worst of all I felt like a hypocrite. Here I was on a soapbox about equality, empowerment, and

164 | Yvonne Winkler

selfcare and yet I secretively soothed my pain every night with a bottle of red.

With Kerry on speed dial and equipped with a solid journaling practice and a dozen quit lit books, I began to prepare for recovery. In my previous attempts to abstain from drinking, I distracted myself with keeping busy as an entrepreneur. However, I noticed that the harder I worked myself to burnout, the more I wanted to drink at the end of the day. And so, for years I swung between overwork and drinking. From workaholic to alcoholic and alcoholic back to workaholic. The problem with being a workaholic in today's world was that it is celebrated. The hustle, the boss babe, and the message that when we don't measure up, with drinking or work, we are failures. I learned that the underpinning of addiction is the shame spiral, the thing which kept me from truly connecting to my authentic self and what I wanted to do with my life. To interrupt this pattern, I did something radically different this time. I quit both.

The moment we sold our house I resigned from my managing director position at eWomenNetwork, put my coaching practice on hold and did not engage in any conferences, sales courses, or business networking, no matter how convenient and inexpensive the offer was. My only priority was to get through another day without drinking alcohol or obsessing over where my next client was coming from.

I started every morning with a meditation, a check-in with a sober group I found online, and a thirty- sometimes sixty-minute walk with Sophie during which I listened to other people's stories of how they got sober. Slowly, I began to start taking joy in the simple things again. Feeding the bunnies on the farm we stayed at over the summer. Reading a novel on the porch with a cold glass of lemonade sweating beside me. I delighted in making simple meals.

One day, while perusing the fresh produce at the local market, I picked up a basket of gleaming red strawberries. To the tune of the grasshoppers chirping in the dry grass outside the kitchen window, I cut them in half and poured fresh milk and a sprinkle of sugar over them. Everything seemed to slow down and come more into focus. Colors looked brighter, the air smelled sweeter, food tasted better. My nervous system was beginning to recover.

As I opened my eyes on the morning of day thirty-three, an unwelcome thought popped into my head. *It's summer, do you really want to be the party pooper on the boat today?* We were invited out onto the sparkling, cool waters of Kalamalka Lake for a Canada Day celebration. Good friends of ours had a permanent summer trailer site nearby, along with a fast boat that was fully decked out with water sport gear and a fantastically loud stereo. Normally I jumped on any opportunity to be hanging out on a boat in the heat of summer with a cold beer in my hand. After a year and a half of not connecting with friends due to COVID restrictions, I was certainly aching to let lose a little, and this group of people knew how to party and enjoy the luxuries of lake country. But how was I going to get through this without drinking when the whole point of this outing was to do just that? *Maybe it's just me who's making a big deal out of this. I'm sure others don't go out boating thinking only about vodka coolers,* I thought.

"What do you want me to pack up for you to drink?" Den came into the bedroom where I was still flopping around with my thoughts.

"Definitely the cucumber mint tonics, they are refreshing and kind of look like those coolers, so it won't be obvious that I'm not drinking." I learned that telling people in advance that I'm not drinking helps take the awkwardness out of the moment by defusing the implication, but we were going to be surrounded by strangers and I didn't want to get into lengthy discussions about it. There were typically two reactions to my not drinking. One, it made the other person so uncomfortable that they would spend a considerable amount of time explaining to me how they can take it or leave it at any time, then not talk to me for the rest of the night because I'm weird. Two, they would ask a lot of questions that I didn't know how to answer yet, still working on my recovery, which inevitably led to them to moving to more fun people to talk to. And so, I preferred to keep to myself about it until such time that I would be able to not care whether my choices met with anyone else's approval. The less it came up, the easier it would be to refrain.

I opened *The Unexpected Joy of Being Sober* and flipped back to something I had read the night before. "Addiction," Catherine Gray wrote, "is not a 'normal drinkers' versus 'alcoholic' division; it's a spectrum. Even drinkers that would be classified as 'normal' in the eyes of a doctor, would find it unimaginable and horrifying to never drink again." Looking at my dependency on a scale helped me understand why my previous attempts to moderate my drinking hadn't worked and that *just one* wasn't a viable option for me anymore. While other people may be able to contain their more than usual drinking to their summer vacation, I couldn't. Although I didn't start drinking with the sunrise, I thought about it all the time. *Did I bring enough wine? Will anyone notice if I pour another? What if we run out?* The cloud around my brain had begun to lift, and life felt so much more stable now. My sleep was getting better, no more hangovers to nurse all day under the disguise of being tired from overwork. I wanted to do whatever it took to stop this endless cycle of self-destruction.

After a day out on the boat, Den and I said goodnight and a big thank you for having been invited. We had a great time. Den even tried wakesurfing, something that wasn't easy for him to do amongst all those lake people who had years of practice on him.

"Wanna go to Pane Vino Pizzeria?" he asked me as we backed out of the trailer site.

"Yes, I'm starving," I replied, glowing from the sun that was still radiating off my skin, but also from having participated in a social function without alcohol.

When the waitress brought us our coconut waters with lemon, I felt overcome with gratitude. I imagined how this day would have played out with me having one of those fancy looking tequila paloma coolers. Instead of shame and regret, I left feeling proud of myself, confidently knowing that I had a good time, and that I didn't make an ass of myself by saying something stupid that might have embarrassed our friends in front of their group. And best of all, I knew I would have a great sleep that night and not feel hungover in the morning.

The two months we spent on the farm in the Okanagan went by quickly, yet I remember almost everything about the days we were there. From the smell of the morning coffee on the porch to the long walks on the dry, dusty path along the turquoise waters of Wood Lake. One of the first things I learned in recovery was to take it one day at a time. Being very intentional and conscious of my behaviors, thoughts, and environment forced me to be in the present moment. The concept wasn't new to me but the implementation and reality of it was. Drinking and overwork were distraction tactics I had developed to not deal with what was in front of me, creating a perpetual existence in either my past or my future. Depression or anxiety, both of which I would numb out with more of the same. This led to my body, mind, and spirit being completely fatigued.

"I wonder if I should make my sobriety journey more public?" I winced into my webcam.

"It's an important part of your story and freedom," Kerry smiled back at me. "Many women struggle with this, and we need to know that we aren't alone."

"I'm scared though," I said, "I'm supposed to be a coach and help them with their freedom lifestyle and creating sustainable work practices, but I can't get my own shit together. What if they think I'm a fraud?" In my head I added, *what if I had to go back to the corporate job?* I knew that showing vulnerability meant a heck of a lot more than a diploma certificate on the wall, yet I was terrified.

"Why don't you play this one by ear. You'll know when it's the right time and what to share," Kerry said empathetically. We had been working together for five years, diligently peeling back, layer by layer, the shield and armor I had constructed over a lifetime of searching for where I belonged. She taught me how to come back to my essential self, the one I was born with before the world told me who I needed to be.

I grabbed my pink faux leather notepad, the one I started on the first day I set foot into Kerry's office to capture important insights. Tucked into the sleeve I saw a quote by American spiritual teacher Adyashanti, given to me by my friend Mini. "Enlightenment is a destructive process. It has nothing to do with becoming better or being happier. Enlightenment is the crumbling away of untruth. It's seeing through the façade of pretense. It's the complete eradication of everything we imagined to be true."

It was destructive and all this reality all the time was so intense. My brain jumped over to the business. What about my brand? What about all the work I had already plowed into this business? I had to make this work, if I didn't I was a huge failure. What would the people I walked away from say? I needed to do more. *Oh, but it's so hard alone,* I agonized. And I was into a full internal tailspin.

Fall, the season of starting a new year was upon us and I could hear the nagging voices come in louder and louder. I had successfully stayed away from obsessing over emails, marketing, funnels, networking, and offerings for the summer by focusing all my energy on not drinking. As people pulled out of their summer retreats and headed back home to get their kids ready for school, we too packed up our truck, left lake country and headed west to our next stop, a tiny laneway house in a quiet neighbourhood of Vancouver's west side, close to the university. The moment we rolled within the city limits I could feel my chest tightening. Raindrops the size of golf balls hit our windshield, the wipers were going as fast as they could to no avail. Den and I looked at each other and immediately knew that our recovery slumber was over. My head sank deeper between my shoulders, and I pulled Sophie closer to my chest. Even she felt the energy shift.

September had always been a very special time of year. It was a transition time and being a seeker, I suppose it made sense that I enjoyed that in-between phase. There was a rustling of possibility in the changing of seasons and a certainty of things coming to rest. I was close to one hundred days alcohol- and work-free, and that meant I was readying myself for the next milestone – one hundred and eighty days.

For my forty-fourth birthday Den and I took a trip to Tofino, a magical place off the west coast of Vancouver Island well-known for its tropical-like sunsets, wild pacific surf, and old growth cedar forest. As we toured along the winding Pacific Rim Highway, Den suggested that we should consider Vancouver Island as a possible landing pad. I hesitated. The island was indeed Canada's version of Hawaii but I was unsure about settling in a small community ever again. Memories of being unwanted and not fitting in flooded my brain. Nova Scotia had left a bitter taste. But, since we had agreed to give every prospective place a try, I parked my hesitation and conceded to look for a place to stay on the more populated east side of the island for the winter, with direct access to the mainland and Vancouver.

During one of our biannual three-hour catch-up conversations Carola once asked, "When is it ever going to be enough for you?" I knew she was referring to my incessant seeking for more, better, something else, to which I replied, "I don't think I'll ever stop looking for my best life."

Walking along the golf course in my new neighborhood, inhaling the salty, humid air as Sophie strutted ahead, I finally realized that what I was seeking lived inside of me. I had no idea that the journey I set out on ten years ago called *back to my roots and beyond* would indeed be about finding the essential parts that made up Yvonne Winkler. I always felt a deeper calling in the pit of my stomach, like I was meant to do great things but I didn't know how. So, I dug myself into everything 1000 percent, and as people told me I would do great things, I dug in more. I looked everywhere, left no stone unturned, and what I discovered is that it couldn't be found outside of me. I already had everything I needed to be who I wanted and the freedom I craved. I was the temple. I had to learn how to take care of it.

There was one fundamental lesson I learned from my parents that day in the ditch, thirty-three years ago. If I can't live with the situation I have, it's up to me to create the future I want. It won't be easy, it won't be comfortable, and it will involve sacrifice. But I am always free to choose.

I had to let go of what wasn't working, once again. My home, drinking, the grip I had on success and independence, my comfort zones and certainty, to make room for what I really wanted, freedom from stifling doubt. Freedom from counting drinks and cigarettes, freedom from reprimanding myself about not being normal or fitting in. Freedom from believing that I could only be of value by minimizing my worth. Freedom from the traps of shame and guilt brought on by not being good enough.

I had to face my past to find my way back to my essential self. What that entailed was learning to forgive; forgive others for knowingly or unknowingly hurting me and then forgive myself for the same. With each prayer of forgiveness I let go of another shackle. That's what I believe the wise sages mean when they say freedom comes from within. It can't be bought, it's not in a place, it's the peace we find within ourselves when we finally recognize that who we are is where we belong.

When I shed the adaptive strategies I had developed to survive the world that I created for myself, I began to radiate my true essence. That, in turn, attracted the people who could love me for who I was, not who they needed me to be. My life began to be alive; my business began to flourish; my relationships began to bloom.

Freedom Seeker | 171

***

I sat in the departure lounge at south gate, waiting for a turboprop to fly me back to Vancouver Island. It was around five in the afternoon, and there were only three other passengers, one security person, and the agent who checked me in. Tired from the two-day training seminar I had attended in Vancouver and nearly two years of doing any type of work online (if at all), I welcomed the quiet retreat and lack of usual commotion at airports.

I took a seat on one of the four empty benches and surveyed my surroundings. The little reception hall resembled a Canadian log cabin and was decorated with pictures of Indigenous art. There was a canoe hanging from the ceiling and a totem to my left. I looked over my right shoulder and spotted a woman wearing a hoody and jeans, stooped over a glass of wine. She sat at one of the five small tables in front of the kiosk that offered coffee in Styrofoam cups with little sugar packets and creamers, right next to the little souvenir shop that displayed the same seal figure my dad had brought back for me from his first trip to Canada. Her face looked weathered by time and smoking, I guessed. That could easily have been me, looking worn from smoking and drinking and fitting in with everyone, wearing the same style clothes. Not paying any attention to my health and wellness because I was more concerned about my next drink.

It'd been eight years since I smoked my last cigarette, over a year since I drank alcohol or drowned myself in debilitating overwork. I took a careful sip of my scalding hot mint tea. My life had slowed down and become simpler, just like my commute back to Vancouver Island. And it was exactly in this simplicity that my life became large. I had once looked for fulfillment in the grandness of things like the Audi, the city, the vacations. But I had found it in the most unexpected place; a small community that hadn't forgotten that it's about *we*, not *me*.

Den and I had to make a choice between continuing our life in a comfortable and predictable pattern or listen to our hearts and leap into a new adventure free from obligation and guilt. We did it together and we did it by choosing ourselves first. Here on the island, in our new home and a community that welcomed us even though we didn't know we were looking for it. I found a new solace with the freedom of routine, starting my days with meditation, followed by a walk with Sophie that ends with a breathtaking view of the mountains in the distance and the familiar cry of a seagull welcoming me home.

AUTHORS NOTE

Most women have learned to be pleasers and appeasers, and many of us operate from a place of scarcity (a.k.a. "I am not enough") simply because we've been ignored for centuries. We don't feel that we belong or that we are enough because patriarchal values and male-centric social customs have demonized feminine values for thousands of years. We've been asked to hide our most essential parts, and we've learned to fit in and play up the masculine values that are not in alignment with our feminine wisdom. We compare and compete, creating only a greater divide and distrust amongst women.

Both energies are equally important.

The feminine is responsible for desire and the masculine for producing that desire.

Ignoring the importance of both qualities has left us lost and seeking validation in things outside ourselves. I watched my mother, my grandmothers, my aunts, and other women in my life prioritize their husbands, families, or careers, working really, really, really, really hard (because it's never enough!), self-sacrificing and ignoring their own needs. Insisting on matching what they were feeling inside about their value with their outer circumstances; burning out, bitter, resentful, and depressed.

When we learn to take care of ourselves first and that we matter equally, we reclaim the source of our power. Only then are we capable of doing what comes naturally to us—taking care of others. The fuller we are, the more generous we can become. And the ripples from our potency spread far and wide, impacting everyone around us.

Throughout the pages of this book, I talk about the Lotus Wellness Temple, which has indeed come into form, albeit different from how I first imagined it on the beach in Tarifa. Through the ups and downs of my personal developmental journey, I know, as certain as the sun rises, that when I take care of my brain, my heart, and my body, life is more enjoyable, my relationships are better, and I am free. I am the temple.

As a student of the Academy of Emerging Women, and through my lived experiences and as a Birkman Certified Professional (BCP), I can get to the heart of the matter and support women in the discovery process of what they really want to do with their precious life. It lights me up to watch my clients break free from stifling self-doubt, counting steps, calories, drinks, cigarettes. Free from the shame of being different, free from loneliness and distrust, and free from dependencies.

I believe that our world needs a harmony of both masculine and feminine energy. It is my intention to help restore this balance by championing feminine values as I coach corporate women and entrepreneurs one-on-one. I've also taken everything I've learned in this book and rallied a community of women passionate about making a difference in the world around them through collaboration. Join the Freedom Seeker Community at yvonnewinkler.com/community/

Finally, see the full color Freedom Seeker book companion photo album with pictures from my childhood, *back to my roots and beyond* trip and life today at yvonnewinkler.com/freedom-seeker-book/

ACKNOWLEDGMENTS

It all began one Tuesday afternoon in my coach Kerry's office, when she proposed that writing my story down for other women to read could quite possibly be the medicine for my aching heart and unrelenting search for purpose and meaning in my life. She saw my essential self when I was still lost in the dark, and her glimmer of light and belief in me was what I needed to begin my path of becoming a writer. Kerry Parsons, without your tireless nudging to look for the wisdom in my heart, I wouldn't have been able to navigate the emotional and often paralyzing journey into the depths of my past. You were one of the keys that helped unlock my freedom.

Brittany Veenhuysen, thank you for being the first person to guide my fingers along the keyboard and publish my early writings.

I'm grateful for Marie Maccagno and her Adventures in Writing group for helping me build my penning confidence and for teaching me how to tap into the flow of wisdom.

Kati Pauls, thank you for visually bringing my work and this book alive through your artful cover and graphic designs.

Deep appreciation to the entire team at Big Sky Author Services for clearly laying out all the steps to becoming a published author. Thank you to my copy editor, Liz Cook, and LeeAnn Lessard for distribution.

A special thank you to my beta readers Heather, Patty, and Kerry. I know firsthand the time involved in reading an unpolished draft and the delicate nature of providing critical feedback. You helped me make this book better.

Thank you to my launch team and affiliate partners. Your support and belief in me are a continuous source of strength.

To my writing coach and developmental editor Tammy Plunkett, words can hardly convey my gratitude for you. From the moment I came to you with this book idea, you dedicated all of yourself to my success. This book would not exist without your wisdom, love, and guidance.

Thank you to my mom and dad, Barbara and Siegfried Winkler, for spending hours on the phone with me reliving the good, the bad, and the ugly of our family's journey, filling in the missing pieces my early childhood brain didn't understand, and fact-checking my historical account of events. Mutti, your strength and perseverance never ceases to amaze me. Vati, thank you for being so wildly courageous and for showing me what's possible with the quest for freedom in our hearts.

To my husband, Dennis French, thank you for all the nourishing meals, dog walks, baths, and gentle reminders to take breaks. The words of encouragement when muse was absent, and for patiently listening to my first, second, and third drafts. I'm so grateful for you as a provider and true partner in life. I love you.

CPSIA information can be obtained
at www.ICGtesting.com
Printed in the USA
LVHW051606251122
733960LV00003B/440